A First Course in Logic

A First Course in Logic

Mark V. Lawson
Heriot-Watt University, Edinburgh

CRC Press
Taylor & Francis Group
Boca Raton London New York

CRC Press is an imprint of the
Taylor & Francis Group, an **informa** business

CRC Press
Taylor & Francis Group
6000 Broken Sound Parkway NW, Suite 300
Boca Raton, FL 33487-2742

© 2019 by Taylor & Francis Group, LLC
CRC Press is an imprint of Taylor & Francis Group, an Informa business

No claim to original U.S. Government works

Printed on acid-free paper

International Standard Book Number-13: 978-0-8153-8665-0 (Paperback)
International Standard Book Number-13: 978-0-8153-8664-3 (Hardback)

Library of Congress Cataloging-in-Publication Data

Names: Lawson, Mark V., author.
Title: A first course in logic / Mark V. Lawson.
Description: Boca Raton, Florida : CRC Press, [2018] | Includes
bibliographical references and index.
Identifiers: LCCN 2018024672| ISBN 9780815386643 (hardback : alk. paper) |
ISBN 9780815386650 (pbk. : alk. paper) | ISBN 9781351175388 (ebook).
Subjects: LCSH: Logic, Symbolic and mathematical--Problems, exercises, etc. | Logic.
Classification: LCC QA9 .L37256 2018 | DDC 511.3--dc23
LC record available at https://lccn.loc.gov/2018024672

Visit the Taylor & Francis Web site at
http://www.taylorandfrancis.com

and the CRC Press Web site at
http://www.crcpress.com

This book is dedicated to the memory of my mother

Shirley Lawson (1935–2017)

"...for the growing good of the world is partly dependent on un-historic acts; and that things are not so ill with you and me as they might have been, is half owing to the number who lived faithfully a hidden life ..." – George Eliot.

Contents

Preface

When you come to a fork in the road, take it! — Yogi Berra.

This book grew out of notes written to accompany my Heriot-Watt University module *F17LP Logic and proof* which I have been teaching since 2011. This module was in fact instigated by my colleagues in Computer Science and was therefore intended originally for first-year computer science students, but the module was subsequently also offered as an option to second-year mathematics students (in the Scottish university system). There are three chapters. Chapters 1 and 2 focus on propositional logic and Boolean algebras, respectively, and Chapter 3 is a short introduction to the basic ideas of first-order logic. I have treated propositional logic as a subject in its own right rather than as just a stopover to a more interesting destination. In particular, I outline the intuitive background to the question of whether $\mathcal{P} = \mathcal{NP}$, describe how this question is related to the satisfiability problem of propositional logic, and why all this matters. The theory of Boolean algebras is the algebraic face of propositional logic and is used to design circuits: both the combinational ones that have no memory and the sequential ones that do. Each section concludes with exercises, all of whose solutions can be found at the end of the book. Let me stress that this is very much a *first* introduction to logic. I have therefore assumed as few prerequisites as possible; specifically, I have tried to keep any mathematical background to the absolute minimum. The real mathematical prerequisite is an ability to manipulate symbols: in other words, basic algebra. Anyone who can write programs should have this ability already. In addition, I do not try to follow every logical byway, rather my goal is to inspire interest and curiosity about this subject and lay the foundations for further study. In writing this book, I have been particularly influenced, like many others, by the work of Smullyan[1], particularly [67]. Chapters 1 and 3 cover roughly the same material as the first 65 pages of his [67] together with part of his Chapter XI. For a complete list of all the references I consulted, see the bibliography, but let me highlight a few particular examples here. In both Chapter 1 and Chapter 3, I have used truth trees. These are vastly superior to the Hilbert-style systems often contained in many introductions to logic. Though such systems have their place, they are a stumbling block in an introduction such as this. I first learnt about them from [65]. I have, however, included a short introduc-

---

[1] Raymond Smullyan (1919–2017) was the doyen of logic, and his work did much to bring it to a wider audience.

tion to sequent calculus at the conclusion of Chapter 1 since this is a system important in more advanced applications, but I have also shown how it is closely related to truth trees. I do not include a formal description of natural deduction, which is natural only to someone already familiar with mathematical proof, but I do touch on natural deduction informally in Section 3.3. The material in Section 2.3 on combinational circuits is unexceptional but I was inspired by Eck [14] to include a specific binary logic model of transistors; given the singular importance of such devices it is surprising that this has not become standard. Sequential circuits tend not to be handled in modern logic books, which is an odd omission. My treatment of such circuits in Section 2.4 was inspired by reading [8]. The material on sequential circuits requires the notion of a finite-state automaton in various guises. Such machines are special kinds of the Turing machines needed in more advanced courses on logic and complexity theory. For this reason, Chapter 1 contains a light introduction to finite-state automata. This enables me to indicate how Cook's theorem, stated in Section 1.9, is actually proved which should render it less enigmatic. I cannot claim any real innovations in my discussion of first-order in Chapter 3 except that I have striven to be as concrete and clear as possible. I deal only with structures having constants and relations though I do allude to structures with functions; this makes life easier and enables me to concentrate on the main goal of that chapter which is a proof of Gödel's Completeness Theorem. In addition, I make no mention of equality in first-order logic which really becomes relevant when logic is applied to the study of mathematical structures. My overriding goal in writing this book was to unite the theoretical — logic as used in mathematics — with the more practical — logic as used in computer science. My perspective is that this demarcation is completely analogous to that between pure and applied mathematics and that both approaches are mathematics.

My thanks to Phil Scott (Ottawa) and Takis Konstantopoulos (Uppsala) for reading and commenting on an early draft. Thanks also to the following philanthropists: Till Tantau for TikZ, Sam Buss for bussproofs, Jeffrey Mark Siskind and Alexis Dimitriadis for qtree and Alain Matthes for tkz-graph. Kudos to the following students for spotting typos: Chi Chan, Alexander Evetts (my TA in 2017), Muhammad Hamza, Dimitris Kamaritis, Andrea Lin and Jessica Painter. Finally, my thanks to colleagues at CRC Press for their support and help: Paul Boyd, Callum Fraser, Cindy Gallardo, Sarfraz Khan and Shashi Kumar.

Errata, etc. I shall post these at the following page also accessible via my homepage:

http://www.macs.hw.ac.uk/~markl/A-first-course-in-logic.html

Mark V. Lawson
Edinburgh, Vernal Equinox and Summer Solstice, 2018

Introduction

> **Logic**, *n. The art of thinking and reasoning in strict accordance with the limitations and incapacities of the human misunderstanding.* — Ambrose Bierce.

Logic is the study of how we reason and, as such, is a subject of interest to philosophers, mathematicians and computer scientists. This book is aimed at mathematicians and computer scientists (but philosophers are welcome as well). The word 'logic' itself is derived from the Greek word 'logos' one of whose myriad meanings is 'reason'.[2] Human beings used reason, when they chose to, thousands of years before anyone thought to study *how* we reason. The formal study of logic only began in classical Greece with the influential work of the philosopher Aristotle and a circle of philosophers known as the Stoics, but the developments significant to this book are more recent beginning in the nineteenth century.[3] The Greek legacy is an important one, and as a result the Greek alphabet is widely used in science and mathematics, and this book is no exception. For reference, I have included a table of the upper- and lowercase Greek letters below.

---

[2] The online version of Liddell and Scott, a famous dictionary of classical Greek, to be found at `http://stephanus.tlg.uci.edu/lsj/#eid=65855&context=lsj&action=from-search`, contains two pages of meanings of the word 'logos' in very small type. The same word, with another of its meanings — 'word' — crops up in the first line of St John's Gospel.

[3] The concept of 'logic' as a discipline was largely confined to academic circles but in the UK, at least, this all changed on a sunny Saturday afternoon on the 12th July 1969 when the BBC began broadcasting *Star Trek* with the episode "Where No Man Has Gone Before" which contained the character of Mr Spock, half-human and half-vulcan played by Leonard Nimoy (1931–2015), who was guided by a philosophy inspired by logic.

Introduction

The Greek Alphabet

α	alpha	A	ν	nu	N
β	beta	B	ξ	xi	Ξ
γ	gamma	Γ	o	omicron	O
δ	delta	Δ	π	pi	Π
ϵ	epsilon	E	ρ	rho	P
ζ	zeta	Z	σ	sigma	Σ
η	eta	H	τ	tau	T
θ	theta	Θ	υ	upsilon	Υ
ι	iota	I	ϕ	phi	Φ
κ	kappa	K	χ	chi	X
λ	lambda	Λ	ψ	psi	Ψ
μ	mu	M	ω	omega	Ω

Two names that stand out as the progenitors of modern logic are George Boole (1815–1864) and Gottlob Frege (1848–1925). Their work, as well as that of others — see the graphic novel [13], for example, or the weightier tome [41] — led in 1928 to the first textbook on mathematical logic *Grundzüge der theoretischen Logik* [30] by David Hilbert (1862–1943) and Wilhelm Ackermann (1896–1962). This served not only as the template for all subsequent introductions to mathematical logic, it also posed a question always referred to in its original German: the *Entscheidungsproblem* (which simply means the *decision problem*). The solution of the *Entscheidungsproblem* by Alan Turing (1912–1954) can, for once, be described without hyperbole as revolutionary.[4]

Turing was one of the most influential mathematicians of the twentieth century; he is often claimed as the father of computer science, a claim that rests partly on the paper he wrote in 1937 [75] entitled *On computable numbers, with an application to the Entscheidungsproblem*. There Turing describes what we now call in his honour a *universal Turing machine*, a mathematical blueprint for building a computer independent of technology.[5] Remarkably, Turing showed in his paper that there were problems that could not be solved by computer — not because the computers weren't powerful enough but because they were intrinsically unable to solve them. Thus the limitations of computers were known before the first one had ever been built. Turing didn't set out to invent computers, rather he set out to solve the *Entscheidungsproblem*, the very problem posed by Hilbert and Ackermann in their book. Roughly speaking, this asks whether it is possible to write a program that will answer all mathematical questions. Turing proved that this was impossible. The *Entscheidungsproblem* is, in fact, a question about logic and it is my hope that

---

[4]The first solution to this problem was by Alonzo Church (1903–1995) using a system he introduced called lambda calculus. But Turing's solution is the more intuitively compelling. Interestingly, whereas Turing's work had an influence on hardware, Church's had an influence on software.

[5]You can access copies of this paper via the links to be found on the Wikipedia article on the *Entscheidungsproblem*. See also [62].

by the end of this book you will understand what is meant by this problem and its significance for logic.

Introductory exercises

The exercises below do not require any prior knowledge but introduce ideas that are important in this book.

1. Here are two puzzles by Raymond Smullyan.[6] On an island there are two kinds of people: *knights* who always tell the truth and *knaves* who always lie. They are indistinguishable.

   (a) You meet three such inhabitants A, B and C. You ask A whether he is a knight or knave. He replies so softly that you cannot make out what he said. You ask B what A said and he says 'he said he is a knave'. At which point C interjects and says 'that's a lie!'. Was C a knight or a knave?

   (b) You encounter three inhabitants: A, B and C.
   A says 'exactly one of us is a knave'.
   B says 'exactly two of us are knaves'.
   C says: 'all of us are knaves'.
   What type is each?

2. This question is a variation of one that has appeared in the puzzle sections of many magazines. There are five houses, from left to right, each of which is painted a different colour, their inhabitants are called Sarah, Charles, Tina, Sam and Mary, *but not necessarily in that order*, who own different pets, drink different drinks and drive different cars.

   (a) Sarah lives in the red house.

   (b) Charles owns the dog.

   (c) Coffee is drunk in the green house.

   (d) Tina drinks tea.

   (e) The green house is immediately to the right (that is: your right) of the white house.

   (f) The Oldsmobile driver owns snails.

   (g) The Bentley owner lives in the yellow house.

   (h) Milk is drunk in the middle house.

   (i) Sam lives in the first house.

---

[6]This question and the next are recycled from [44].

(j) The person who drives the Chevy lives in the house next to the person with the fox.

(k) The Bentley owner lives in a house next to the house where the horse is kept.

(l) The Lotus owner drinks orange juice.

(m) Mary drives the Porsche.

(n) Sam lives next to the blue house.

There are two questions: who drinks water and who owns the aardvark?

3. *Bulls and Cows* is a code-breaking game for two players: the code-setter and the code-breaker. The code-setter writes down a four-digit secret number all of whose digits must be different. The code-breaker tries to guess this number. For each guess they make, the code-setter scores their answer: for each digit in the right position score 1 bull (1B), for each correct digit in the wrong position score 1 cow (1C); no digit is scored twice. The goal is to guess the secret number in the smallest number of guesses. For example, if the secret number is 4271, and I guess 1234, then my score will be $1B, 2C$. Here's an easy problem. The following is a table of guesses and scores. What are the possibilities for the secret number?

1389	$0B, 0C$
1234	$0B, 2C$
1759	$1B, 1C$
1785	$2B, 0C$

4. *Hofstadter's MU-puzzle.*[7] A *string* is just an ordered sequence of symbols. In this puzzle, you will construct strings using the letters M, I, U where each letter can be used any number of times, or not at all. You are given the string MI which is your only input. You can make new strings from old only by using the following rules any number of times in succession in any order:

(I) If you have a string that ends in I then you can add a U on at the end.

(II) If you have a string Mx where x is a string then you may form Mxx.

(III) If III occurs in a string then you may make a new string with III replaced by U.

(IV) If UU occurs in a string then you may erase it.

I shall write $x \to y$ to mean that y is the string obtained from the string x by applying one of the above four rules. Here are some examples:

---

[7]This question is adapted from the book [33]. Not uncontroversial, it has the great virtue of being *interesting*. Read it and be inspired!

- By rule (I), $MI \to MIU$.
- By rule (II), $MIU \to MIUIU$.
- By rule (III), $UMIIIMU \to UMUMU$.
- By rule (IV), $MUUUII \to MUII$.

The question is: can you make MU?

5. *Sudoku puzzles* have become very popular in recent years. The newspaper that first launched them in the UK went to great pains to explain that they had nothing to do with mathematics despite involving numbers. Instead, they said, they were logic problems. This of course is nonsense; logic is part of mathematics. What they should have said is that they had nothing to do with *arithmetic*. The goal is to insert digits in the boxes to satisfy two conditions: first, each row and each column must contain all the digits from 1 to 9 exactly once, and second, each 3×3 box, demarcated by bold lines, must contain the digits 1 to 9 exactly once.

	1		4	2				5
		2	7	1		3	9	
						4		
2		7	1					6
				4				
6				7	4			3
	7							
1	2		7	3		5		
3			8	2		7		

Chapter 1

Propositional logic

The world is all that is the case. — Ludwig Wittgenstein.

The main goal of this book is to introduce *first-order logic* (FOL). But ever since Hilbert and Ackermann's book [30], this has almost always been done by first describing the simpler logic called *propositional logic* (PL), which is the subject of this chapter, and then upgrading to full FOL, which we do in Chapter 3. The logics we discuss are examples of *artificial languages* to be contrasted with *natural languages* such as English. Natural languages are often imprecise when we try to use them in situations where precision is essential, being riddled with ambiguities.[1] In such cases, artificial languages are used. Thus, programming languages are artificial languages suitable for describing algorithms to be implemented by computer. The description of an artificial language has two aspects: *syntax* and *semantics*. Syntax, or grammar, tells you what the allowable sequences of symbols are, whereas semantics tells you what they mean.

1.1 Informal propositional logic

Our goal is to construct a precise, unambiguous language that will enable us to *calculate* the truth or falsity of a sentence in that language. Before defining this precise artificial language, we begin by analysing everyday natural language.

Language consists of sentences, but not all sentences will be grist to our mill. Here are some examples.

1. Homer Simpson is prime minister.

2. The Earth orbits the Sun.

3. To be or not to be?

4. Out, damned spot!

---

[1] The word 'cleave', for example, means both stick together and split apart.

Sentences (1) and (2) are different from sentences (3) and (4) in that we can say of sentences (1) and (2) whether they are true or false — in this case, (1) is false[2] and (2) is true — whereas for sentences (3) and (4) it is meaningless to ask whether they are true or false, since (3) is a question and (4) is an exclamation.

> A sentence that is capable of being either true or false (though we might not know which) is called a *statement*.[3]

In science, it is enough to study only statements[4] which come in all shapes and sizes from the banal 'the sky is blue' to the informative 'the Battle of Hastings was in 1066'. But in our work, the only thing that interests us about a statement is whether it is *true (T)* or *false (F)*. In other words, what its *truth value* is and nothing else; the phrase 'and nothing else' is important. We can now give some idea as to what the subject of logic is about:

> Logic is about deducing whether a given statement is true or false on the basis of information provided by some (other) collection of statements.

I shall say more about this later.

Some statements can be analysed into combinations of simpler statements using special kinds of words called *(propositional) connectives*. The connectives we shall be interested in are **not**, **and**, **or**, **xor**, **iff** (this is not a typo), and **implies** and are described below. I have written them in bold to indicate that they will sometimes be used in unusual ways. I should add that these are not the only connectives; there are in fact infinitely many of them. In Section 1.6, we shall make precise the sense in which the ones above are more than sufficient.

not

Let p be the statement *It is raining*. This is related to the statement *It is not raining*, which we shall denote by q, because the truth value of q will be the opposite of the truth value of p. This is because q is the *negation* of p. This is precisely described by the following truth table:

It is raining	It is not raining
T	F
F	T

To avoid peculiarities of English grammar, we replace the word 'not' in the

---

[2]Satire is outwith the remit of logic.

[3]Or, a *proposition*. This is why 'propositional logic' is so called. But as a mathematician, I use the word 'proposition' to mean a theorem of lesser importance. I was therefore uncomfortable using it in the sense needed in this book.

[4]If we did this in everyday life, we would come across as robotic.

first instance by the slightly less natural phrase 'It is not the case that'. Thus *It is not the case that it is raining* means the same thing as *It is not raining.*[5] We go one step further and abbreviate the phrase 'It is not the case that' by **not**. Thus if we denote a statement by p then its negation is **not** p. The above table becomes:

p	**not** p
T	F
F	T

This summarizes the behaviour of negation for *any* statement p and not just those dealing with meteorology. What happens if we negate twice? Then we simply apply the above table twice to obtain:

p	**not not** p
T	T
F	F

Thus *It is not the case that it is not raining* should mean the same as *It is raining*. I know this sounds weird and it would certainly be an odd thing to say as anything other than a joke but we are building here a language suitable for science in general, and mathematics and computer science in particular, rather than everyday usage.

The word **not** is our first example of a connective. It is also what is called a *unary* connective because it is only applied to one input. The remaining connectives are called *binary* because they are applied to two inputs.

and

Under what circumstances is the statement *It is raining and it is cold* true? Well, I had better be both wet and cold. However, in everyday English the word 'and' often means 'and then'. The statement *I got up and had a shower* does not mean the same thing as *I had a shower and got up*. The latter sentence might be a joke: perhaps my roof leaked when I was asleep. Our goal is to eradicate the ambiguities of everyday language so we certainly cannot allow these two meanings to coexist. Therefore in logic only the first meaning is the one we use. To make it clear that I am using the word 'and' in a special sense, I shall write it in bold: like this **and**. Given two statements p and q, we can describe the truth values of the compound statement p **and** q in terms of the truth values assumed by p and q by means of the following truth table:

p	q	p **and** q
T	T	T
T	F	F
F	T	F
F	F	F

---

[5]If you used that phrase in everyday language you would sound odd, possibly like a lawyer.

This table tells us that the statement *Homer Simpson is prime minister* **and** *the Earth orbits the Sun* is false. In everyday life, we might struggle to know how to interpret such a sentence — if someone turned to you on the bus and said it, I think your response would be one of alarm rather than to register that it was false. Let me underline that the *only* meaning we attribute to the word **and** is the one described by the above truth table. Thus contrary to everyday life, the statements *I got up* **and** *I had a shower* and *I had a shower* **and** *I got up* mean the same thing.

or and xor

The word *or* in English is more troublesome. Imagine the following setup. You have built a voice-activated robot that can recognize shapes and colours. It is placed in front of a white square, a black square and a black circle.

You tell the robot to choose a black shape or a square. It chooses the black square. Is that good or bad? Well, it depends what you mean by 'or'; it can mean *inclusive or*, in which case the choice is good, or it can mean *exclusive or*, in which case the choice is bad. Both meanings are useful, so rather than choose one over the other we use two different words to cover the two different meanings. This is an example of *disambiguation*. We use the word **or** to mean *inclusive or*. This is defined by the following truth table:

p	q	p **or** q
T	T	T
T	F	T
F	T	T
F	F	F

Thus p **or** q is true when *at least one* of p and q is true. We use the word **xor** to mean *exclusive or*. This is defined by the following truth table:

p	q	p **xor** q
T	T	F
T	F	T
F	T	T
F	F	F

Thus p **xor** q is true when *exactly one* of p and q is true. Although we haven't got far into our study of logic, this is already a valuable exercise. If you use the word *or*, you should know what you really mean.

iff

Our next propositional connective will be familiar to students of mathematics but less so to anyone else. This is the connective *is equivalent to*, also expressed as *if and only if*, and usually written as **iff**, which is an abbreviation and not evidence of the author's atrocious spelling, and different from the word 'if'. This is defined by the following truth table:

p	q	p **iff** q
T	T	T
T	F	F
F	T	F
F	F	T

Observe that **not**$(p\,$**iff**$\,q)$ means the same thing as $p\,$**xor**$\,q$. This is our first result in logic. By the way, I have added brackets to make it clear that we are negating the whole statement $p\,$**iff**$\,q$ and not merely p.

We use brackets in logic to clarify our meaning. They are not mere decoration.

implies

Our final connective is problematical: *if p then q* which we shall write as $p\,$**implies**$\,q$. Much has been written about this connective — its definition may even go back to the Stoics mentioned in the Introduction — because it unfailingly causes problems. In order to understand it, you have to remember that binary connectives simply join statements together and that the statements can be anything at all. In particular, there need be no connection whatsoever between them. In everyday language, this is unnatural because we acquire rules from the society around us on how the different parts of a sentence should be related to each other.[6] Clearly, we cannot write these complex social rules into our formal language, quite apart from the fact that in science we are simply not interested in statements that have social nuance. Thus, whatever meaning we attribute to '$p\,$**implies**$\,q$' it must hold for *any* choices of p and q. In addition, we have to be able to say what the truth value of the compound statement '$p\,$**implies**$\,q$' is in terms of the truth values of p and q. To figure out what the truth table for this connective should be, we consider it from two different points of view.

1. **Implies** *as a promise.* Suppose your parents say the following 'If you pass your driving test, then we shall buy you an E-Type Jaguar'.[7] If you

---

[6]What these rules are belongs to a subject called *pragmatics*. Getting them wrong, however perfect your pronunciation or flawless your grammar, can be the source of humour or conflict. As an example, consider the following situation. Parent says to teenager: what time do you call this? Teenager replies: midnight.

[7]A British car widely viewed as the most beautiful ever built.

do indeed pass your driving test and they do indeed buy you a Jaguar, then they will have fulfilled their promise. On the other hand, if you pass your driving test and they don't buy you a Jaguar then you would legitimately complain that they had broken their promise. Both of these cases are easy. Suppose, however, you don't pass your driving test. If they don't buy you a Jaguar, you cannot complain since they certainly haven't broken their promise. If they do, nevertheless, buy you a Jaguar, then you would be delighted since it would be completely unexpected. But, again, they wouldn't be breaking the terms of their promise (they just weren't telling you the whole story). In summary, the promise is kept in all situations except where you pass your test and your parents don't buy you that E-Type Jaguar.

2. **Implies** *has to fit in with the other connectives.* Suppose I tell you that the truth value of the statement p is T and that the truth value of the whole statement 'p **implies** q' is also T. What can we say about the truth value of q? Intuitively, we would say that q also has the truth value T. This enables us to complete the first row of the truth table below:

p	q	p implies q
T	T	T
T	F	x
F	T	y
F	F	z

where x, y and z are truth values we have yet to determine. Assume that p is T and q is F. If the truth value for $p \to q$ were T then by row one, we would have to deduce that q was T, as well. This conflicts with our assumption that q is F. It follows that x must be F. Next, p **iff** q should mean the same thing as $(p$ **implies** $q)$ **and** $(q$ **implies** $p)$. A quick calculation shows that this forces z to be T. We have therefore filled in three rows of the truth table below:

p	q	p implies q
T	T	T
T	F	F
F	T	y
F	F	T

Finally, we are left with having to decide whether y is T or F. If we chose F then we would have a truth table identical to that of **iff**. But **implies** should be different from **iff**; thus we are forced to put $y = T$.

This gives us the following truth table for **implies**:

p	q	p implies q
T	T	T
T	F	F
F	T	T
F	F	T

It should be noted that the definition of **implies** we have given does have some seemingly bizarre consequences when used in everyday situations. For example, the statement *Homer Simpson is prime minister* **implies** *the Sun orbits the Earth* is in fact true. This sort of thing can be offputting when first encountered and can seem to undermine what we are trying to achieve. But remember: we are using everyday words in very special ways. The key to all of this is the following:

> As long as we translate between English and logic, choosing the correct words to reflect the meaning we intend to convey, then everything will be fine.

I have used the bold symbols **not**, **and**, etc., above. But these are rooted in the English language. It would be preferable to use a notation which is independent of a specific language since our logic will be a universal language. The table below shows what we shall use in this book.

English	Symbol	Name
not	$\neg$	negation
and	$\wedge$	conjunction
or	$\vee$	disjunction
xor	$\oplus$	exclusive disjunction
iff	$\leftrightarrow$	biconditional
implies	$\rightarrow$	conditional

Our symbol for **and** is essentially a capital 'A' with the crossbar missing, and our symbol for **or** is the first letter of the Latin word *vel* which means 'or'.

A statement that cannot be analysed further using the propositional connectives is called an *atomic statement* or simply an *atom*. Otherwise a statement is said to be *compound*. The truth value of a compound statement can be determined once the truth values of its atoms are known by applying the truth tables of the propositional connectives defined above.

Examples 1.1.1. Determine the truth values of the following statements.

1. (There is a rhino under the table) $\vee \neg$(there is a rhino under the table). *Always true.*

2. $(1 + 1 = 3) \to (2 + 2 = 5)$. *True.*

3. (Mickey Mouse is the president of the USA) $\leftrightarrow$ (pigs can fly). *Amazingly, true.*

Example 1.1.2. Consider a simple traffic control system that consists of a red light and a green light. If the light is green traffic flows and if the light is red traffic stops. Let g be the statement 'the light is green', let r be the statement 'the light is red' and let s be the statement 'the traffic stops'. Then the following compound statements are all true: $r \oplus g$, $g \to \neg s$ and $r \to s$. Suppose I tell you that g is true. Then from the fact that $r \oplus g$ is true, we deduce that r is false. Now $g \to \neg s$ is true and g is true and so $\neg s$ is true. Thus the traffic doesn't stop. Since $\neg s$ is true it follows that s is false. Now observe that in $r \to s$ both r and s are false, but we are told that the statement $r \to s$ is true. This is entirely consistent with the way that we have defined $\to$.

Example 1.1.3. My car has all kinds of warnings both audible and visual.[8] For example, 'the audible warning for headlamps sounds if the key is removed from the ignition and the driver's door is open and either the headlamps are on or the parking lamps are on'. Put p equal to the statement 'the key is removed from the ignition', put q equal to 'the driver's door is open', put r equal to 'the headlamps are on' and put s equal to 'the parking lamps are on'. Put t equal to 'the audible warning for the headlamps sounds'. Then t is true if $(p \wedge q) \wedge (r \vee s)$ is true. Observe that $r \vee s$ is the correct version of 'or' since we need at least the headlamps to be on or the parking lamps to be on, but that should certainly include the possibility that both sets of 'lamps' are on. Although the guidebook uses the word 'if', I think it is clear that it really means 'if and only if'. Thus we want t to be true precisely when $(p \wedge q) \wedge (r \vee s)$ is true. For example, I don't want the audible warning for the headlamps to sound if the 'gas' is low. Thus if $(p \wedge q) \wedge (r \vee s)$ is true the audible warning for the headlamps sounds and if $(p \wedge q) \wedge (r \vee s)$ is false, then it doesn't.

What we have done so far is informal in that I have just highlighted some features of everyday language. What we shall do next is formalize. I shall describe to you an artificial language, called PL, motivated by what we have described in this section. I shall first describe its syntax and then its semantics. Of course, I haven't shown you yet what we can actually do with this artificial language. That I shall do later.

---

[8]The guidebook for my car is written in American English but the steering wheel is on the right side.

Exercises 1.1

---

1. For each of the following statements, decide whether they are true or false.

    (a) The Earth orbits the Sun.

    (b) The Moon orbits Mars.

    (c) Mars orbits the Sun.

    (d) Mars orbits Jupiter.

    (e) (The Moon orbits Mars) ↔ (Mars orbits Jupiter).

    (f) (The Earth orbits the Sun) → (Mars orbits the Sun).

    (g) (The Earth orbits the Sun) ⊕ (Mars orbits the Sun).

    (h) (The Earth orbits the Sun) ⊕ (Mars orbits Jupiter).

    (i) (The Earth orbits the Sun) ∨ (Mars orbits the Sun).

    (j) (The Earth orbits the Sun) ∨ ¬(the Earth orbits the Sun).

2. The *Wason selection task*. Below are four cards, each of which has a number printed on one side and a letter on the other.

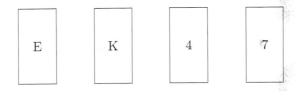

    The following claim is made: if a card has a vowel on one side, then it has an even number on the other. What is the smallest number of cards you have to turn over to verify this claim and which cards are they?

1.2 Syntax of propositional logic

On the basis of our informal lucubrations in Section 1.1, we now define the formal language called *propositional logic* (PL). This consists of a set of symbols called *atomic statements* or *atoms* and denoted by $p_1, p_2, p_3, \ldots$ (though we shall usually use lowercase letters $p, q, r, \ldots$ for convenience) from which we construct *well-formed formulae* (wff) defined as follows:

(WFF1). All atoms are wff.

(WFF2). If X and Y are wff then so too are $(\neg X)$, $(X \wedge Y)$, $(X \vee Y)$, $(X \oplus Y)$, $(X \to Y)$ and $(X \leftrightarrow Y)$.

(WFF3). All wff are constructed by repeated application of the rules (WFF1) and (WFF2) a finite number of times.

A wff which is not an atom is said to be a *compound statement*.

Example 1.2.1. We show that

$$(\neg((p \vee q) \wedge r))$$

is a wff in accordance with the definition above in a sequence of steps:

1. p, q and r are wff by (WFF1).

2. $(p \vee q)$ is a wff by (1) and (WFF2).

3. $((p \vee q) \wedge r)$ is a wff by (1), (2) and (WFF2).

4. $(\neg((p \vee q) \wedge r))$ is a wff by (3) and (WFF2), as required.

Notational convention. To make the reading of wff easier for humans, I shall omit the final outer brackets and also the brackets associated with $\neg$.

Examples 1.2.2.

1. $\neg p \vee q$ means $((\neg p) \vee q)$.

2. $\neg p \to (q \vee r)$ means $((\neg p) \to (q \vee r))$ and is different from $\neg((p \to q) \vee r)$.

I tend to bracket fairly heavily but many books on logic (and in this one, occasionally) use fewer brackets by applying a convention. First, the connectives are arranged in the following hierarchy

$$\neg, \wedge, \vee, \to, \leftrightarrow$$

which should be read from left to right with $\neg$ as top dog, all the way down to $\leftrightarrow$ as the runt of the litter. This hierarchy is interpreted as follows: if two connectives are next to each other in a wff, you apply first the one that is higher up the food chain. This is analogous to what we do in elementary arithmetic. For example, $a + b \cdot c$ means $a + (b \cdot c)$ because multiplication has higher precedence than addition. Likewise, the expression $p \to q \wedge r$ should be interpreted as the wff $p \to (q \wedge r)$ because $\wedge$ has precedence over $\to$.

Trees

There is a graphical way of representing wff that uses what are called trees. A *tree* is a data structure which will play an important role in this book in a number of different ways; we shall always think of them pictorially. They consist of *vertices*, which we represent by circles, and *edges*, which we represent by lines joining two different vertices. The vertices are arranged in horizontal levels which we could number $0, 1, 2, \ldots$. At the top, at level 0, is one vertex called the *root*. Next come the vertices at level 1, then the vertices of level 2, and so on. A vertex u and a vertex v can only be joined by an edge if the level of u and the level of v differ by exactly 1. The vertices at the lowest level, if there are any, are called *leaves*. Finally, it is convenient to draw the tree 'upside-down' with the root at the top; later, we shall deal with trees that are upside-down versions of these and so will be arboreally correct in having their roots at the bottom. The picture below is an example of a tree with the leaves being indicated by the black circles. The root is the vertex at the top:

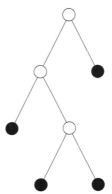

In the following tree, I have numbered the vertices for convenience:

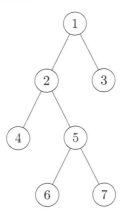

We say that vertices 2 and 3 are the *successors* of vertex 1. Likewise, vertices 4 and 5 are the successors of vertex 2.

Looked at the other way around, vertex 1 is the *predecessor* of vertices 2 and 3. Likewise, vertex 2 is the predecessor of vertices 4 and 5. A *path* in a tree begins at the root and then follows the edges down to another vertex. Thus $1 - 2 - 5$ is a path. A path that ends at a leaf is called a *branch*. Thus $1 - 2 - 5 - 7$ is a branch. A branch is therefore a maximal path. Finally, a *descendant w* of a vertex v is a vertex that can be obtained from v by following a sequence of successors to v. Thus any successor of v is a descendant of v and any successor of a descendant of v is a descendant of v.

Parse trees

We use this notion of a tree to define the following. A *parse tree of a wff* is constructed as follows. The parse tree of an atom p is the tree

Now let A and B be wff. Suppose that A has parse tree T_A and B has parse tree T_B. Let $*$ denote any of the binary propositional connectives. Then $A * B$ has the parse tree

This is constructed by joining the roots of T_A and T_B to a new root labelled by $*$. The parse tree for $\neg A$ is

This is constructed by joining the root of T_A to a new root labelled $\neg$. Parse trees are a way of representing wff without using brackets, though we pay the price of having to work in two dimensions rather than one.

We shall see in Chapter 2 that parse trees are in fact useful in circuit design.

Example 1.2.3. The parse tree for $\neg(p \vee q) \vee r$ is

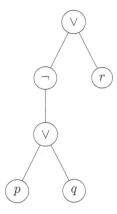

Example 1.2.4. The parse tree for $\neg\neg((p \rightarrow p) \leftrightarrow p)$ is

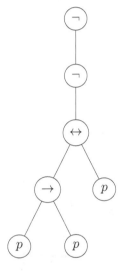

Although we shall not need this definition now, it is convenient to introduce a rather obvious notion of the complexity of a wff called its *degree*; this is simply the total number of connectives in the wff. Thus, atoms have degree 0 and the wff of degree 1 are those wff that contain exactly one connective.

I conclude this section with a twofold mathematical aside that will enable us to say precisely, in what sense, PL is a language.

Sets

Set theory is a sort of *lingua franca* in mathematics and computer science, and you will need a smattering to read this book. It was invented by Georg Cantor (1845–1918) in the last quarter of the nineteenth century and begins with two deceptively simple definitions on which everything is based:

1. A *set*[9] is a collection of objects, called *elements*, which we wish to view as a whole.

2. Two sets are *equal* precisely when they contain the same elements.

To make clear what is in the set we use curly brackets { and } (officially these are called 'braces') to mark the boundaries of the set. So, for example, the set of suits in a pack of cards is $\{\clubsuit, \spadesuit, \heartsuit, \diamondsuit\}$. However, there doesn't have to be any rhyme or reason to what we put into a set. For example, I typed the following set $\{\star, \P, \nabla, \blacksquare, \eth\}$ at random. A set is a single thing so it is usual to give it a name. What you call a set doesn't matter in principle, but in practice we use uppercase letters such as $A, B, C \ldots$ or fancy uppercase letters such as $\mathbb{N}, \mathbb{Z} \ldots$ with the elements of the set usually being denoted by lowercase letters. The special symbol $\varnothing$ is used to name the *empty set*, {}, which otherwise might leave you wondering whether I had forgotten something. I could, therefore, write

$$S = \{\clubsuit, \spadesuit, \heartsuit, \diamondsuit\}$$

where the set S has as elements $\clubsuit$, $\spadesuit$, $\heartsuit$ and $\diamondsuit$. If 'x *is an element of* the set A' then we write '$x \in A$' and if 'x *is not an element of* the set A' then we write '$x \notin A$.' A set is *finite* if it only has a finite number of elements, otherwise it is *infinite*. The definition of equality for sets has two very important, and not entirely sensible-looking, consequences:

1. A set should be regarded as a bag of elements, not a display case, and so *the order of the elements within the set is not important*; thus $\{a, b\} = \{b, a\}$.

2. More surprisingly, *repetition of elements is ignored*. Thus $\{a, b\} = \{a, a, a, b, b, a\}$. Sets can be generalized to what are called *multisets* where repetition is recorded but we shall not need this concept in this book.

To check that $A = B$, two statements have to be verified:

1. If $a \in A$ then $a \in B$.

2. If $a \in B$ then $a \in A$.

This is the basis of the formal method used in mathematics for showing that two sets are equal. A set with exactly one element is called a *singleton set*. Thus $\{a\}$ is an example of a singleton set. The element of this singleton set is a.

---

[9]Be warned that mathematicians like to embellish their language with all kinds of synonyms for the plain word 'set': collection, aggregate, ensemble, space, class . . . .

Observe that $a \neq \{a\}$ because the left-hand side is an element and the right-hand side is a set that contains one element: namely, a.[10]

Example 1.2.5. On the face of it, sets seem simple sorts of critters but things take a complex turn when you realise that a set can have elements which are themselves sets. For example

$$X = \{a, \{b\}, \{a, \{b\}\}, \{\{a\}\}\}$$

is a set with four elements, three of which are themselves sets. It is in this way that sets become complicated and, indeed, complicated enough that all of mathematics can be carried out using them.

Strings

Let A be a finite set. In certain circumstances, we can also call this set an *alphabet*.[11] This term is often used when the elements of a set are going to be used to form 'strings' often with the goal of being used in communication. A *string* over the alphabet A is an ordered sequence of elements from A. For example, if $A = \{0, 1\}$ then the following are examples of strings over this alphabet: 0, 1, 00, 01, 10, 11,.... Observe that strings are usually written by simply *concatenating* the elements of the alphabet; that is, by placing the elements of the alphabet next to each other in order. The *empty string* is the string that contains no elements of the alphabet and is denoted by ε. The total number of elements of the alphabet occurring in a string is called its *length*. Thus 00000 has length 5 and ε has length 0. Strings over the alphabet $\{0, 1\}$ are usually called *binary strings*. Given two strings x and y, we can stick them together to get a new string xy. This operation is called *concatenation*. Clearly, order matters in applying this operation. The concatenation of the strings 'go' and 'od' is the string 'good' whereas the concatenation of the strings 'od' and 'go' is 'odgo'. A set of strings is called, suggestively, a *language*.

---

[10]If I have a sheep as a pet, I could hardly be said to have a flock of sheep, the word 'flock' being used instead of the word 'set' when dealing with sheep collections. On the other hand, if I do have a flock of sheep but, sadly, all bar one of the sheep were eaten by ravenous wolves then I do have a flock, albeit consisting of one lone sheep. Incidently, English has the odd feature that it uses different words to mean set or collection in different situations. Thus a herd is a set of cows, a school is a set of whales and a murmuration is a set of starlings.

[11]I already have a footnote warning you about the propensity mathematicians have for ransacking Roget's Thesaurus for words that are synonyms for the word 'set'. Part of this is style but mainly, I think, it is psychological; we use that synonym for the word 'set' that gives us a sense of what kind of set it is supposed to be in a particular context.

Example 1.2.6. We now use the above terminology to describe what we have done in this section. Our starting point is the (infinite) alphabet

$$A = \{\neg, \wedge, \vee, \oplus, \rightarrow, \leftrightarrow, (,), p_1, p_2, p_3, \ldots\};$$

observe that the ... tells us to carry on in the same way. We are not interested in all strings over this alphabet but only those strings constructed in accordance with the three rules (WFF1), (WFF2) and (WFF3). We call the set of such strings the *language of well-formed formulae*.

Exercises 1.2

1. Construct parse trees for the following wff.

   (a) $(\neg p \vee q) \leftrightarrow (q \rightarrow p)$.

   (b) $p \rightarrow ((q \rightarrow r) \rightarrow ((p \rightarrow q) \rightarrow (p \rightarrow r)))$.

   (c) $(p \rightarrow \neg p) \leftrightarrow \neg p$.

   (d) $\neg(p \rightarrow \neg p)$.

   (e) $(p \rightarrow (q \rightarrow r)) \leftrightarrow ((p \wedge q) \rightarrow r)$.

2. Let $A = \{\clubsuit, \diamondsuit, \heartsuit, \spadesuit\}$, $B = \{\spadesuit, \diamondsuit, \clubsuit, \heartsuit\}$ and $C = \{\spadesuit, \diamondsuit, \clubsuit, \heartsuit, \clubsuit, \diamondsuit, \heartsuit, \spadesuit\}$. Is it true or false that $A = B$ and $B = C$? Explain.

3. Write down how many elements each of the following sets contains.

   (a) $\varnothing$.

   (b) $\{\varnothing\}$.

   (c) $\{\varnothing, \{\varnothing\}\}$.

   (d) $\{\varnothing, \{\varnothing\}, \{\varnothing, \{\varnothing\}\}\}$.

1.3   Semantics of propositional logic

In Section 1.2, we defined the formal language PL. Specifically, we defined well-formed formulae (wff). In this section, we shall assign meanings to such wff; this is what we mean by the word 'semantics'. An atom has exactly one of two *truth values*: either *true (T)* or *false (F)*. If an atom is a specific statement such as 'the earth orbits the sun' then we can say exactly what its truth value is: in this case it is T. However, if an atom is a symbol such as p then all we know is that it is either T or F, but not which. We now consider the truth values of those compound statements that contain exactly one of the propositional connectives. The following six *truth tables* **define** the meaning of the propositional connectives:

p	$\neg p$
T	F
F	T

p	q	$p \wedge q$
T	T	T
T	F	F
F	T	F
F	F	F

p	q	$p \vee q$
T	T	T
T	F	T
F	T	T
F	F	F

p	q	$p \leftrightarrow q$
T	T	T
T	F	F
F	T	F
F	F	T

p	q	$p \oplus q$
T	T	F
T	F	T
F	T	T
F	F	F

p	q	$p \rightarrow q$
T	T	T
T	F	F
F	T	T
F	F	T

As we discussed in Section 1.1, the meanings of the propositional connectives above are *suggested* by their meanings in everyday language, but are not the same as them. Think of our definitions as technical ones for technical purposes only.

> *It is vitally important in what follows that you learn the above truth tables by heart and be able to justify them at a moment's notice.*

Example 1.3.1. The above truth tables tell us how to calculate truth values of wff involving one propositional connective when we know the truth values of the atoms. For example, consider the wff $p \rightarrow q$. Suppose that p is the statement 'the Earth orbits the Sun' and q is the statement 'gold is less dense than lead'. Then we know that p is true and q is false. Thus from the truth table for $\rightarrow$, the compound statement 'the Earth orbits the Sun implies gold is less dense than lead' is false.

We can now calculate the truth values of all compound statements. This is accomplished by using more general truth tables to list all the possibilities. Let A be a compound statement consisting of the atoms $p_1, \ldots, p_n$. A specific *truth assignment* to $p_1, \ldots, p_n$ leads to a specific truth value being assigned to A itself by repeatedly using the definitions above. The following example illustrates this.

Example 1.3.2. Let $A = (p \vee q) \rightarrow (r \leftrightarrow \neg s)$. A truth assignment is given by the following table

p	q	r	s
T	F	F	T

If we insert these values into our wff we get the expression

$$(T \vee F) \to (F \leftrightarrow \neg T).$$

We now use the truth tables above to evaluate this expression in stages. First, $T \vee F$ evaluates to T and $\neg T$ evaluates to F so that our expression simplifies to the new expression

$$T \to (F \leftrightarrow F).$$

Next, $F \leftrightarrow F$ evaluates to T and so the above expression simplifies to

$$T \to T.$$

Finally, $T \to T$ evaluates to T. As a result of these calculations, it follows that the truth value of A for the above truth assignment is T.

The proof of the following lemma uses a technique known as *proof by induction*. This is a technique we shall use in a few places in this book. An explanation as to why it works will be given in Section 2.1.

Lemma 1.3.3. *If the statement X has exactly n atoms then the truth table for X has exactly 2^n rows.*

Proof. The number of rows of the truth table is equal to the number of strings of length n over the alphabet $\{T, F\}$; see the conclusion to Section 1.2 for the terminology I have used. Induction arguments always come in two parts. First part: suppose that $n = 1$. Then there are exactly two strings over the alphabet $\{T, F\}$ of length 1: namely, T and F. But $2 = 2^1$. It follows that the result holds when $n = 1$. Second part: assume that there are 2^k strings over the alphabet $\{T, F\}$ of length k. I shall prove that, as a consequence of this assumption, there are 2^{k+1} strings over the alphabet $\{T, F\}$ of length $k + 1$. Let x be such a string. Then $x = Ty$ or $x = Fy$ where y is a string length k over the alphabet $\{T, F\}$. It follows that there are twice as many strings of length $k + 1$ over the alphabet $\{T, F\}$ as there are strings of length k over the alphabet $\{T, F\}$. But there are 2^k strings of length k over the alphabet $\{T, F\}$ by assumption. Thus there are $2 \cdot 2^k = 2^{k+1}$ strings of length $k + 1$ over the alphabet. The proof of the lemma now follows from these two parts by induction. $\square$

Example 1.3.4. Lemma 1.3.3 is less innocuous than it looks since it is an example of 'exponential explosion'. This is illustrated by the following story. There was once a king who commissioned a fabulous palace from an architect. So pleased was the king with the result, that he offered the architect as much gold as he could carry from the royal treasury. The architect said that just seeing his work completed was wealth enough but that if the king was really insistent on rewarding him, he would be content with just the gold coins that could be placed on a chessboard in the following way: one gold coin should be placed on the first square, two gold coins on the second square, four gold coins on the third square and so on. The king, not known for his mathematical acumen, readily agreed thinking this a small price to pay. In what way was the king duped by the architect?[12]

For each wff A, we may therefore draw up a truth table whose rows consist of all possible truth assignments to the atoms of A along with the corresponding truth value of A. We shall use the following pattern of assignments of truth values:

...	T	T	T
...	T	T	F
...	T	F	T
...	T	F	F
...	F	T	T
...	F	T	F
...	F	F	T
...	F	F	F
...	...	...	...

Thus for the rightmost column, truth values alternate starting with T; for the next column to the left, they alternate in pairs starting with TT; for the next column to the left, they alternate in fours starting with $TTTT$; and so on.

Examples 1.3.5. Here are some examples of truth tables.

1. The truth table for $A = \neg(p \to (p \vee q))$ is

p	q	$\neg(p \to \neg(p \vee q))$
T	T	F
T	F	F
F	T	F
F	F	F

However, it is a bad idea to carry out truth table calculations in your head since it is all too easy to make mistakes. I will therefore present the truth table for A with extra columns representing intermediate calculations. In this case, we would have the following

p	q	$p \vee q$	$p \to (p \vee q)$	A
T	T	T	T	F
T	F	T	T	F
F	T	T	T	F
F	F	F	T	F

---

[12]Just to complete the story, the king did eventually realize his mistake, but being an enlightened monarch his vengeance was mild and the architect was merely required to spend the rest of his career designing houses people actually wanted to live in.

2. The truth table for $B = (p \wedge (p \rightarrow q)) \rightarrow q$, showing intermediate calculations, is

p	q	$p \rightarrow q$	$p \wedge (p \rightarrow q)$	B
T	T	T	T	T
T	F	F	F	T
F	T	T	F	T
F	F	T	F	T

3. The truth table for $C = (p \vee q) \wedge \neg r$ is

p	q	r	$p \vee q$	$\neg r$	C
T	T	T	T	F	F
T	T	F	T	T	T
T	F	T	T	F	F
T	F	F	T	T	T
F	T	T	T	F	F
F	T	F	T	T	T
F	F	T	F	F	F
F	F	F	F	T	F

4. Given the wff $(p \wedge \neg q) \wedge r$, we could draw up a truth table but in this case we can easily figure out how it behaves. It is true if and only if p is true and $\neg q$ is true and r is true. Thus the following is the truth assignment that makes the wff true

p	q	r
T	F	T

and the wff is false for all other truth assignments. We shall generalize this example later.

Important definitions

- An atom or the negation of an atom is called a *literal*.[13]

- We say that a wff A built up from the atoms $p_1, \ldots, p_n$ is *satisfiable* if there is some assignment of truth values to its atoms which gives A the truth value true.

- If $A_1, \ldots, A_n$ are wff we say they are *(jointly) satisfiable* if there is a single truth assignment that makes all of $A_1, \ldots, A_n$ true.

- If a wff is always true we say that it is a *tautology*. If A is a tautology we shall write

$$\vDash A.$$

The symbol $\vDash$ is called the *semantic turnstile*.

---

[13]I have no explanation as to why this term is used.

- If a wff is always false we say it is a *contradiction*. If A is a contradiction we shall write

$$A \vDash .$$

Observe that contradictions are on the left (or sinister) side of the semantic turnstile.

- If a wff is sometimes true and sometimes false we refer to it as a *contingency*.

- A truth assignment that makes a wff true is said to *satisfy* the wff, otherwise it is said to *falsify* the wff.

A very important problem in PL can now be stated.

The satisfiability problem (SAT)

Given a wff in PL, decide whether there is some truth assignment to the atoms that makes the wff take the value true. A program that solves SAT is called a *SAT solver*.

I shall discuss SAT in more detail later and explain why it is significant.

It is important to remember that all questions in PL can be settled, at least *in principle*, by using truth tables.

Example 1.3.6. What's the problem with truth tables? Why did I say above *in principle*? This goes back to Example 1.3.4, which we now look at in a more mathematical way. Suppose you want to construct the truth table of a wff that has 90 atoms (not a vast number by any means). Its truth table will therefore have 2^{90} rows. It is more useful to express this as a power of 10. You can check that $2 \approx 10^{0.3}$, where the symbol $\approx$ means *approximately equal to*. Thus

$$2^{90} \approx (10^{0.3})^{90} = 10^{0.3 \times 90} \approx 10^{27}.$$

For the sake of argument, suppose that it takes you 10^{-9} seconds to construct each row of the truth table. Then it will take you $10^{-9} \times 10^{27} = 10^{18}$ seconds to construct the whole truth table. For comparison purposes, the age of the universe, give or take a few seconds, is 4.35×10^{17} seconds.[14]

The logical constants

We conclude with a very mild addition to PL which is, nevertheless, often useful. Let f be an 'atom' that is always false and t be an 'atom' that is always true. We call f and t *logical constants*; they are connectives that don't connect anything. This addition to PL does not change its expressive power one iota but it often enables us to state results in much more memorable forms, as we shall see.

---
[14]What is this in millions of years?

Exercises 1.3

1. Determine which of the following wff are satisfiable. For those which are, find all the assignments of truth values to the atoms which make the wff true.

   (a) $(p \wedge \neg q) \rightarrow \neg r$.

   (b) $(p \vee q) \rightarrow ((p \wedge q) \vee q)$.

   (c) $(p \wedge q \wedge r) \vee (p \wedge q \wedge \neg r) \vee (\neg p \wedge \neg q \wedge \neg r)$.

2. Determine which of the following wff are tautologies by using truth tables.

   (a) $(\neg p \vee q) \leftrightarrow (q \rightarrow p)$.

   (b) $p \rightarrow ((q \rightarrow r) \rightarrow ((p \rightarrow q) \rightarrow (p \rightarrow r)))$.

   (c) $(p \rightarrow \neg p) \leftrightarrow \neg p$.

   (d) $\neg(p \rightarrow \neg p)$.

   (e) $(p \rightarrow (q \rightarrow r)) \leftrightarrow ((p \wedge q) \rightarrow r)$.

3. We defined only 5 binary connectives, but there are in fact 16 possible ones. The tables below show all of them.

p	q	$\circ_1$	$\circ_2$	$\circ_3$	$\circ_4$	$\circ_5$	$\circ_6$	$\circ_7$	$\circ_8$
T	T	T	T	T	T	T	T	T	T
T	F	T	T	T	T	F	F	F	F
F	T	T	T	F	F	T	T	F	F
F	F	T	F	T	F	T	F	T	F

p	q	$\circ_9$	$\circ_{10}$	$\circ_{11}$	$\circ_{12}$	$\circ_{13}$	$\circ_{14}$	$\circ_{15}$	$\circ_{16}$
T	T	F	F	F	F	F	F	F	F
T	F	T	T	T	T	F	F	F	F
F	T	T	T	F	F	T	T	F	F
F	F	T	F	T	F	T	F	T	F

Express each of the connectives from 1 to 16 in terms of at most $\neg$, $\wedge$, $\vee$, $\rightarrow$, $\leftrightarrow$, $\oplus$, p, q and brackets. There are many correct solutions to this question. You can check your answers using the truth table generator [76].

1.4 Logical equivalence

It can happen that two different-looking statements A and B have the same truth table. This means they have the same meaning. For example, 'it is not the case that it is not raining' is just a bizarre way of saying 'it is raining'. In this section, we shall investigate this idea.

1.4.1 Definition

We begin with some examples.

Example 1.4.1. Compare the true tables of $p \rightarrow q$ and $\neg p \vee q$.

p	q	$p \rightarrow q$
T	T	T
T	F	F
F	T	T
F	F	T

p	q	$\neg p$	$\neg p \vee q$
T	T	F	T
T	F	F	F
F	T	T	T
F	F	T	T

They are clearly the same.

Example 1.4.2. Compare the true tables of $p \leftrightarrow q$ and $(p \rightarrow q) \wedge (q \rightarrow p)$.

p	q	$p \leftrightarrow q$
T	T	T
T	F	F
F	T	F
F	F	T

p	q	$p \rightarrow q$	$q \rightarrow p$	$(p \rightarrow q) \wedge (q \rightarrow p)$
T	T	T	T	T
T	F	F	T	F
F	T	T	F	F
F	F	T	T	T

They are clearly the same.

Example 1.4.3. Compare the truth tables of $p \oplus q$ and $(p \vee q) \wedge \neg(p \wedge q)$.

p	q	$p \oplus q$
T	T	F
T	F	T
F	T	T
F	F	F

p	q	$p \vee q$	$p \wedge q$	$\neg(p \wedge q)$	$(p \vee q) \wedge \neg(p \wedge q)$
T	T	T	T	F	F
T	F	T	F	T	T
F	T	T	F	T	T
F	F	F	F	T	F

They are clearly the same.

If the wff A and B have the same truth tables, we say that A is *logically equivalent to B* written $A \equiv B$.

It is important to remember that $\equiv$ is not a logical connective. It is a *relation* between wff.

The above examples can now be expressed in the following way.

Examples 1.4.4.

1. $p \to q \equiv \neg p \vee q$.

2. $p \leftrightarrow q \equiv (p \to q) \wedge (q \to p)$.

3. $p \oplus q \equiv (p \vee q) \wedge \neg(p \wedge q)$.

However, there is an extra point to bear in mind: A and B need not have the same atoms but, in that case, the truth tables must be constructed using all the atoms that occur in either A or B. Here is an example.

Example 1.4.5. We prove that $p \equiv p \wedge (q \vee \neg q)$. We construct two truth tables with atoms p and q in both cases.

p	q	p		p	q	$p \wedge (q \vee \neg q)$
T	T	T		T	T	T
T	F	T		T	F	T
F	T	F		F	T	F
F	F	F		F	F	F

The two truth tables are the same and so the two wff are logically equivalent.

The following result is the first indication of the important role that tautologies play in propositional logic. You can also take this as the *definition* of logical equivalence since it sidesteps the slight problem in the case where the two wff use different numbers of atoms.

Proposition 1.4.6. *Let A and B be wff. Then $A \equiv B$ if and only if $\vDash A \leftrightarrow B$.*

Proof. The statement of the result is in fact two statements in one since this is the meaning of the phrase 'if and only if':

1. If $A \equiv B$ then $\vDash A \leftrightarrow B$.

2. If $\vDash A \leftrightarrow B$ then $A \equiv B$.

Let the atoms that occur in either A or B be $p_1, \ldots, p_n$. We now prove the two statements.

(1) Let $A \equiv B$ and suppose that $A \leftrightarrow B$ were not a tautology. Then there is an assignment of truth values to the atoms $p_1, \ldots, p_n$ such that A and B have different truth values. But this would imply that there was a row of the truth table of A that was different from the corresponding row of B. This contradicts the fact that A and B have the same truth tables. It follows that $A \leftrightarrow B$ is a tautology.

(2) Let $A \leftrightarrow B$ be a tautology and suppose that A and B have truth

tables that differ. This implies that there is a row of the truth table of A that is different from the corresponding row of B. Thus there is an assignment of truth values to the atoms $p_1, \ldots, p_n$ such that A and B have different truth values. But this would imply that $A \leftrightarrow B$ is not a tautology. $\qquad \square$

Example 1.4.7. The following truth table proves that $\vDash p \leftrightarrow (p \wedge (q \vee \neg q))$.

p	q	$p \wedge (q \vee \neg q)$	$p \leftrightarrow (p \wedge (q \vee \neg q))$
T	T	T	T
T	F	T	T
F	T	F	T
F	F	F	T

It therefore follows that $p \equiv p \wedge (q \vee \neg q)$.

The notion of logical equivalence also provides us with a first opportunity to use the logical constants introduced in the last section. Observe that the wff A such that $A \equiv t$ are precisely the tautologies and the wff A such that $A \equiv f$ are precisely the contradictions.

1.4.2 Logical patterns

As you study PL, you will find that certain logical equivalences constantly recur. The following theorem lists the most important ones. You will be asked to prove them all in the exercises to this section.

Theorem 1.4.8.

1. $\neg \neg p \equiv p$. *Double negation.*

2. $p \wedge p \equiv p$ and $p \vee p \equiv p$. *Idempotence.*

3. $(p \wedge q) \wedge r \equiv p \wedge (q \wedge r)$ and $(p \vee q) \vee r \equiv p \vee (q \vee r)$. *Associativity.*

4. $p \wedge q \equiv q \wedge p$ and $p \vee q \equiv q \vee p$. *Commutativity.*

5. $p \wedge (q \vee r) \equiv (p \wedge q) \vee (p \wedge r)$ and $p \vee (q \wedge r) \equiv (p \vee q) \wedge (p \vee r)$. *Distributivity.*

6. $\neg(p \wedge q) \equiv \neg p \vee \neg q$ and $\neg(p \vee q) \equiv \neg p \wedge \neg q$. *De Morgan's laws.*

7. $p \vee (p \wedge q) \equiv p$ and $p \wedge (p \vee q) \equiv p$. *Absorption.*

> *I will refer to the above results by name in many subsequent calculations, so it is important to learn them all now.*

There are some interesting patterns in the above results that involve the interplay between $\wedge$ and $\vee$:

$p \wedge p \equiv p$	$p \vee p \equiv p$
$(p \wedge q) \wedge r \equiv p \wedge (q \wedge r)$	$(p \vee q) \vee r \equiv p \vee (q \vee r)$
$p \wedge q \equiv q \wedge p$	$p \vee q \equiv q \vee p$

and

$p \wedge (q \vee r) \equiv (p \wedge q) \vee (p \wedge r)$	$p \vee (q \wedge r) \equiv (p \vee q) \wedge (p \vee r)$
$\neg(p \wedge q) \equiv \neg p \vee \neg q$	$\neg(p \vee q) \equiv \neg p \wedge \neg q$
$p \vee (p \wedge q) \equiv p$	$p \wedge (p \vee q) \equiv p$

There are also some important consequences of the above theorem.

- The fact that $(p \wedge q) \wedge r \equiv p \wedge (q \wedge r)$ means that we can write simply $p \wedge q \wedge r$ without ambiguity because the two ways of bracketing this expression lead to the same truth table. It can be shown that as a result we can write expressions like $p_1 \wedge p_2 \wedge p_3 \wedge p_4$ (and so on) without brackets because it can be proved that however we bracket such an expression leads to the same truth table. What we have said for $\wedge$ also applies to $\vee$. The proof of these two claims follows from a (tricky) argument to be found in [44, Box 4.1].

- Let $A_1, \ldots, A_n$ be wff. We abbreviate

$$A_1 \wedge A_2 \wedge \ldots \wedge A_n$$

by

$$\bigwedge_{i=1}^{n} A_i.$$

Similarly, we abbreviate

$$A_1 \vee A_2 \vee \ldots \vee A_n$$

by

$$\bigvee_{i=1}^{n} A_i.$$

- Don't be perturbed by apparently complex-looking expressions such as $\bigwedge_{i=1}^{m} \bigwedge_{j=1}^{n} A_{ij}$. This simply means that you should take the conjunction of all expressions of the form A_{ij} for all possible values of i and j. There will therefore be mn such expressions from A_{11} to A_{mn}. If the i and the j are *both* required to take all possible values between 1 and m, then we might write $\bigwedge_{1 \le i,j \le m} A_{ij}$ rather than $\bigwedge_{i=1}^{m} \bigwedge_{j=1}^{m} A_{ij}$.

- The fact that $p \wedge q \equiv q \wedge p$ implies that the order in which we carry out a sequence of conjunctions does not matter. What we have said for $\wedge$ also applies to $\vee$. The proof of these two claims follows from a (tricky) argument to be found in [44, Box 4.2].

- The fact $p \wedge p \equiv p$ means that we can eliminate repeats in conjunctions of one and the same atom. What we have said for $\wedge$ also applies to $\vee$.

- De Morgan's laws can be generalized. First of all, we have that

$$\neg\left(\bigvee_{i=1}^{n} A_i\right) \equiv \bigwedge_{i=1}^{n} \neg A_i \text{ and } \neg\left(\bigwedge_{i=1}^{n} A_i\right) \equiv \bigvee_{i=1}^{n} \neg A_i.$$

Next, we can *push negation through brackets*. For example,

$$\neg(p \vee (q \wedge r)) \equiv \neg p \wedge \neg(q \wedge r) \equiv \neg p \wedge (\neg q \vee \neg r)$$

by repeated application of De Morgan's laws.

Example 1.4.9. Using associativity and commutativity, we have that

$$p \wedge q \wedge p \wedge q \wedge p \equiv p \wedge q.$$

However, it is important not to overgeneralize as the following two examples show.

Example 1.4.10. Observe that $p \to q \not\equiv q \to p$ since the truth assignment

p	q
T	F

makes the left-hand side equal to F but the right-hand side equal to T.

Example 1.4.11. Observe that $(p \to q) \to r \not\equiv p \to (q \to r)$ since the truth assignment

p	q	r
F	F	F

makes the left-hand side equal to F but the right-hand side equal to T.

1.4.3 Applications

Our next example is an application of some of our results.

Example 1.4.12. We have defined a binary propositional connective $\oplus$ such that $p \oplus q$ is true when exactly one of p or q is true. Our goal now is to extend this to three atoms. Define $\mathrm{xor}(p, q, r)$ to be true when exactly one of p, q or r is true, and false in all other cases. We can describe this connective in terms of the ones already defined. Specifically,

$$\mathrm{xor}(p, q, r) = (p \vee q \vee r) \wedge \neg(p \wedge q) \wedge \neg(p \wedge r) \wedge \neg(q \wedge r).$$

This can easily be verified by constructing the truth table of the right-hand side. Put

$$A = (p \vee q \vee r) \wedge \neg(p \wedge q) \wedge \neg(p \wedge r) \wedge \neg(q \wedge r).$$

p	q	r	$p \vee q \vee r$	$\neg(p \wedge q)$	$\neg(p \wedge r)$	$\neg(q \wedge r)$	A
T	T	T	T	F	F	F	F
T	T	F	T	F	T	T	F
T	F	T	T	T	F	T	F
T	F	F	T	T	T	T	T
F	T	T	T	T	T	F	F
F	T	F	T	T	T	T	T
F	F	T	T	T	T	T	T
F	F	F	F	T	T	T	F

This example is generalized in Question 5 of Exercises 1.4.

The following properties of logical equivalence will be important when we come to show how Boolean algebras are related to PL in Chapter 2.

Proposition 1.4.13. *Let A, B and C be wff.*

1. $A \equiv A$.

2. *If $A \equiv B$ then $B \equiv A$.*

3. *If $A \equiv B$ and $B \equiv C$ then $A \equiv C$.*

4. *If $A \equiv B$ then $\neg A \equiv \neg B$.*

5. *If $A \equiv B$ and $C \equiv D$ then $A \wedge C \equiv B \wedge D$.*

6. *If $A \equiv B$ and $C \equiv D$ then $A \vee C \equiv B \vee D$.*

Proof. By way of an example, I shall prove (6). We are given that $A \equiv B$ and $C \equiv D$ and we have to prove that $A \vee C \equiv B \vee D$. That is, we need to prove that from $\vDash A \leftrightarrow B$ and $\vDash C \leftrightarrow D$ we can deduce $\vDash (A \vee C) \leftrightarrow (B \vee D)$. Suppose that $(A \vee C) \leftrightarrow (B \vee D)$ were not a tautology. Then there would be some truth assignment to the atoms that would make $A \vee C$ true and $B \vee D$ false or vice versa. I shall just deal with the first case here. Suppose that $A \vee C$ were true and that $B \vee D$ were false. Then both B and D would be false and at least one of A and C would be true. If A were true then this would contradict $A \equiv B$, and if C were true then this would contradict $C \equiv D$. It follows that $A \vee C \equiv B \vee D$, as required. $\square$

The above proposition also implies that logical equivalence can be used to *simplify* compound statements, where by 'simplify' we mean 'reduce the degree of'; that is, 'reduce the number of connectives'. To see how, let A be a compound statement which contains occurrences of the wff X. Suppose that $X \equiv Y$ where Y is simpler than X. Let A' be the same as A except that some or all occurrences of X are replaced by Y. Then it follows by Proposition 1.4.13, though this requires some work, that $A' \equiv A$. Observe, however, that A' is simpler than A.

Example 1.4.14. Let

$$A = p \wedge (q \vee \neg q) \wedge q \wedge (r \vee \neg r) \wedge r \wedge (p \vee \neg p).$$

But

$$p \wedge (q \vee \neg q) \equiv p \text{ and } q \wedge (r \vee \neg r) \equiv q \text{ and } r \wedge (p \vee \neg p) \equiv r$$

and so

$$A \equiv p \wedge q \wedge r.$$

Using truth tables to establish logical equivalences is routine but often onerous. Another approach to solving such problems is to use known logical equivalences to prove new ones without having to construct truth tables.

Examples 1.4.15. Here are some examples of using known logical equivalences to show that two wff are logically equivalent.

1. We show that $p \to q \equiv \neg q \to \neg p$.

$$
\begin{aligned}
\neg q \to \neg p \quad &\equiv \quad \neg\neg q \vee \neg p \text{ by Example 4.1(1)} \\
&\equiv \quad q \vee \neg p \text{ by double negation} \\
&\equiv \quad \neg p \vee q \text{ by commutativity} \\
&\equiv \quad p \to q \text{ by Example 4.1(1)}.
\end{aligned}
$$

2. We show that $(p \to q) \to q \equiv p \vee q$.

$$
\begin{aligned}
(p \to q) \to q \quad &\equiv \quad \neg(p \to q) \vee q \text{ by Example 4.1(1)} \\
&\equiv \quad \neg(\neg p \vee q) \vee q \text{ by Example 4.1(1)} \\
&\equiv \quad (\neg\neg p \wedge \neg q) \vee q \text{ by De Morgan} \\
&\equiv \quad (p \wedge \neg q) \vee q \text{ by double negation} \\
&\equiv \quad (p \vee q) \wedge (\neg q \vee q) \text{ by distributivity} \\
&\equiv \quad p \vee q \text{ since } \vDash \neg q \vee q.
\end{aligned}
$$

3. We show that $p \to (q \to r) \equiv (p \wedge q) \to r$.

$$
\begin{aligned}
p \to (q \to r) \quad &\equiv \quad \neg p \vee (q \to r) \text{ by Example 4.1(1)} \\
&\equiv \quad \neg p \vee (\neg q \vee r) \text{ by Example 4.1(1)} \\
&\equiv \quad (\neg p \vee \neg q) \vee r \text{ by associativity} \\
&\equiv \quad \neg(p \wedge q) \vee r \text{ by De Morgan} \\
&\equiv \quad (p \wedge q) \to r \text{ by Example 4.1(1)}.
\end{aligned}
$$

4. We show that $p \to (q \to r) \equiv q \to (p \to r)$.

$$
\begin{aligned}
p \to (q \to r) \;\; &\equiv \;\; \neg p \lor (q \to r) \text{ by Example 4.1(1)} \\
&\equiv \;\; \neg p \lor (\neg q \lor r) \text{ by Example 4.1(1)} \\
&\equiv \;\; (\neg p \lor \neg q) \lor r \text{ by associativity} \\
&\equiv \;\; (\neg q \lor \neg p) \lor r \text{ by commutativity} \\
&\equiv \;\; \neg q \lor (\neg p \lor r) \text{ by associativity} \\
&\equiv \;\; \neg q \lor (p \to r) \text{ by Example 4.1(1)} \\
&\equiv \;\; q \to (p \to r) \text{ by Example 4.1(1).}
\end{aligned}
$$

5. We show that $(p \to q) \land (p \to r) \equiv p \to (q \land r)$.

$$
\begin{aligned}
(p \to q) \land (p \to r) \;\; &\equiv \;\; (\neg p \lor q) \land (\neg p \lor r) \text{ by Example 4.1(1)} \\
&\equiv \;\; \neg p \lor (q \land r) \text{ by distributivity} \\
&\equiv \;\; p \to (q \land r) \text{ by Example 4.1(1).}
\end{aligned}
$$

6. We show that $\vDash p \to (q \to p)$ by using logical equivalences.

$$
\begin{aligned}
p \to (q \to p) \;\; &\equiv \;\; \neg p \lor (\neg q \lor p) \text{ by Example 4.1(1)} \\
&\equiv \;\; (\neg p \lor p) \lor \neg q \text{ by associativity and commutativity} \\
&\equiv \;\; t \text{ since } \vDash \neg p \lor p.
\end{aligned}
$$

Exercises 1.4

---

1. Prove the following logical equivalences using truth tables.

   (a) $\neg\neg p \equiv p$. Double negation.

   (b) $p \land p \equiv p$ and $p \lor p \equiv p$. Idempotence.

   (c) $(p \land q) \land r \equiv p \land (q \land r)$ and $(p \lor q) \lor r \equiv p \lor (q \lor r)$. Associativity.

   (d) $p \land q \equiv q \land p$ and $p \lor q \equiv q \lor p$. Commutativity.

   (e) $p \land (q \lor r) \equiv (p \land q) \lor (p \land r)$ and $p \lor (q \land r) \equiv (p \lor q) \land (p \lor r)$. Distributivity.

   (f) $\neg(p \land q) \equiv \neg p \lor \neg q$ and $\neg(p \lor q) \equiv \neg p \land \neg q$. De Morgan's laws.

2. Prove the following equivalences using truth tables.

   (a) $p \lor \neg p \equiv t$.

   (b) $p \land \neg p \equiv f$.

   (c) $p \lor f \equiv p$.

(d) $p \vee t \equiv t$.

(e) $p \wedge f \equiv f$.

(f) $p \wedge t \equiv p$.

(g) $p \rightarrow f \equiv \neg p$.

(h) $t \rightarrow p \equiv p$.

3. Prove that $p \oplus (q \oplus r) \equiv (p \oplus q) \oplus r$ using truth tables.

4. Prove the following by using known logical equivalences (rather than using truth tables).

(a) $(p \rightarrow q) \wedge (p \vee q) \equiv q$.

(b) $(p \wedge q) \rightarrow r \equiv (p \rightarrow r) \vee (q \rightarrow r)$.

(c) $p \rightarrow (q \vee r) \equiv (p \rightarrow q) \vee (p \rightarrow r)$.

5. (a) Use [76] to construct the truth table for the following wff.

$$A(p,q,r,s) = (p \vee q \vee r \vee s) \wedge \neg(p \wedge q) \wedge \neg(p \wedge r) \wedge \neg(p \wedge s)$$
$$\wedge \ (q \wedge r) \wedge \neg(q \wedge s) \wedge \neg(r \wedge s).$$

Describe in words the meaning of $A(p,q,r,s)$.

(b) Generalize the construction above to define a similar wff of the form $A(p_1, \ldots, p_n)$, where $n \geq 3$ is arbitrary.

6. This question is a proof of the *Duality Theorem* which makes precise the parallels between $\wedge$ and $\vee$. Note that we shall only deal with wff constructed using the connectives $\neg, \vee, \wedge$ in this question. Let A be any such wff with atoms $p_1, \ldots, p_n$. Denote by A^* the wff obtained from A by replacing every occurrence of $\wedge$ in A by $\vee$, and every occurrence of $\vee$ in A by $\wedge$.

(a) Prove that $A^* \equiv \neg A(\neg p_1, \ldots, \neg p_n)$, where $A(\neg p_1, \ldots, \neg p_n)$ is the wff A in which each occurrence of an atom is negated.

(b) Prove that $\vDash A \leftrightarrow B$ if and only if $\vDash A^* \leftrightarrow B^*$.

1.5 PL in action

Once you have absorbed the basic definitions and mastered the notation, you might well experience a sinking feeling: is that it? PL can seem like a toy and some of our examples don't exactly help this impression, but in fact it has serious applications independently of its being the foundation of the more general first-order logic that we shall study in Chapter 3. In this section, we shall describe two applications of PL.

It will be convenient to use generalized xor as described in Question 5 of Exercises 1.4.

1.5.1 PL as a programming language

In this section, we shall describe an example in which PL is used as a sort of programming language. We do this by analyzing a couple of examples of simplified Sudoku-type problems in terms of PL. These illustrate the ideas needed to analyse full Sudoku in terms of PL which is set as an exercise. In fact, many important problems in mathematics and computer science can be regarded as instances of the satisfiability problem. We shall say more about this in Section 1.9.

To understand new ideas always start with the simplest examples. So, here is a childishly simple Sudoku puzzle. Consider the following grid:

where the small squares are called *cells*. The puzzle requires each cell to be filled with a number according to the following two constraints:

(C1) Each cell contains exactly one of the numbers 1 or 2.

(C2) If two cells occur in the same row then the numbers they contain must be different.

There are obviously exactly two solutions to this puzzle

$$\boxed{1\,|\,2} \text{ and } \boxed{2\,|\,1}$$

I shall now show how this puzzle can be encoded by a wff of PL. Please note that I shall encode it in a way that generalizes, so I do not claim that the encoding described in this case is the simplest. We first have to decide what the atoms are. To define them, we shall label the cells as follows:

$$\boxed{c_{11}\,|\,c_{12}}$$

We need four atoms that are defined as follows:

- p is the statement that cell c_{11} contains the number 1.

- q is the statement that cell c_{11} contains the number 2.

- r is the statement that cell c_{12} contains the number 1.

- s is the statement that cell c_{12} contains the number 2.

For example, if p is true then the grid looks like this

$$\boxed{1\,|\,?}$$

where the ? indicates that we don't care what is there. Consider now the following wff.

$$A = (p \oplus q) \wedge (r \oplus s) \wedge (p \oplus r) \wedge (q \oplus s).$$

I now describe what each of the parts of this wff are doing.

- $p \oplus q$ is true precisely when cell c_{11} contains a 1 or a 2 but not both.

- $r \oplus s$ is true precisely when cell c_{12} contains a 1 or a 2 but not both.

- $p \oplus r$ is true precisely when the number 1 occurs in exactly one of the cells c_{11} and c_{12}.

- $q \oplus s$ is true precisely when the number 2 occurs in exactly one of the cells c_{11} and c_{12}.

Here is the important consequence of all of this:

It follows that A is satisfiable precisely when the puzzle can be solved. In addition, each satisfying truth assignment can be used to read off a solution to the original puzzle.

Here is the truth table for A.

p	q	r	s	A
T	T	T	T	F
T	T	T	F	F
T	T	F	T	F
T	T	F	F	F
T	F	T	T	F
T	F	T	F	F
T	F	F	T	T
T	F	F	F	F
F	T	T	T	F
F	T	T	F	T
F	T	F	T	F
F	T	F	F	F
F	F	T	T	F
F	F	T	F	F
F	F	F	T	F
F	F	F	F	F

We observe first that the wff A is satisfiable and so the original problem can be solved. Second, here are the two satisfying truth assignments.

p	q	r	s
T	F	F	T
F	T	T	F

The first truth assignment tells us that $c_{11} = 1$ and $c_{12} = 2$, whereas the second truth assignment tells us that $c_{11} = 2$ and $c_{12} = 1$. These are, of course, the two solutions we saw earlier.

We now step up a gear and consider a more complex, but still simple, example that has more of the flavour of real Sudoku. Consider the following slightly larger grid:

3		
	2	

where again the small squares are called *cells*. This time, some cells contain numbers which must not be changed. Our task is to fill the remaining cells with numbers according to the following constraints:

(C1) Each cell contains exactly one of the numbers 1 or 2 or 3.

(C2) If two cells occur in the same row, then the numbers they contain must be different.

(C3) If two cells occur in the same column, then the numbers they contain must be different.

It is very easy to solve this problem satisfying these constraints to obtain

3	2	1
1	3	2
2	1	3

I shall now show how this problem can be represented in PL and how its solution is a special case of the satisfiability problem. First of all, I shall label the cells in the grid as follows:

c_{11}	c_{12}	c_{13}
c_{21}	c_{22}	c_{23}
c_{31}	c_{32}	c_{33}

The label c_{ij} refers to the cell in row i and column j. PL requires atomic statements, and to model this problem we shall need 27 such atomic statements c_{ijk} where $1 \leq i \leq 3$ and $1 \leq j \leq 3$ and $1 \leq k \leq 3$. The atomic statement c_{ijk} is defined as follows:

c_{ijk} = the cell in row i and column j contains the number k.

For example, the atomic statement c_{113} is true when the grid is as follows:

3	?	?
?	?	?
?	?	?

where the ?'s mean that we don't care what is in that cell. In the above case, the atomic statements c_{111} and c_{112} are both false.

We shall now construct a wff A from the above 27 atoms such that A is satisfiable if and only if the above problem can be solved and such that a satisfying truth assignment can be used to read off a solution. I shall construct A in stages.

- Define $I = c_{113} \land c_{232}$. This wff is true precisely when the grid looks like this

3	?	?
?	?	2
?	?	?

- Each cell must contain exactly one of the numbers $1, 2, 3$. For each $1 \le i \le 3$ and $1 \le j \le 3$, the wff

$$\text{xor}(c_{ij1}, c_{ij2}, c_{ij3})$$

is true when the cell in row i and column j contains exactly one of the numbers $1, 2, 3$. Put B equal to the conjunction of all of these wff. Thus

$$B = \bigwedge_{i=1}^{i=3} \bigwedge_{j=1}^{j=3} \text{xor}(c_{ij1}, c_{ij2}, c_{ij3})$$

where the notation means that you take the conjunction of all the wff $\text{xor}(c_{ij1}, c_{ij2}, c_{ij3})$ for all possible values of i and j where $1 \le i \le 3$ and $1 \le j \le 3$. Thus there are nine terms to be conjoined from $\text{xor}(c_{111}, c_{112}, c_{113})$ through to $\text{xor}(c_{331}, c_{332}, c_{333})$. Then B is true precisely when each cell of the grid contains exactly one of the numbers $1, 2, 3$.

- In each row, each of the numbers $1, 2, 3$ must occur exactly once. For each $1 \le i \le 3$, define

$$R_i = \text{xor}(c_{i11}, c_{i21}, c_{i31}) \land \text{xor}(c_{i12}, c_{i22}, c_{i32}) \land \text{xor}(c_{i13}, c_{i23}, c_{i33}).$$

Then R_i is true when each of the numbers $1, 2, 3$ occurs exactly once in the cells in row i. Define $R = \bigwedge_{i=1}^{i=3} R_i$.

- In each column, each of the numbers $1, 2, 3$ must occur exactly once. For each $1 \le j \le 3$, define

$$C_j = \text{xor}(c_{1j1}, c_{2j1}, c_{3j1}) \land \text{xor}(c_{1j2}, c_{2j2}, c_{3j2}) \land \text{xor}(c_{1j3}, c_{2j3}, c_{3j3}).$$

Then C_j is true when each of the numbers $1, 2, 3$ occurs exactly once in the cells in column j. Define $C = \bigwedge_{j=1}^{i=3} C_i$.

- Put $A = I \wedge B \wedge R \wedge C$. Then by construction A is satisfiable precisely when the original problem is satisfiable and a satisfying truth assignment to the atoms can be used to read off a solution as follows. Precisely the following atoms are true:

$$c_{113}, c_{122}, c_{131}, c_{211}, c_{223}, c_{232}, c_{312}, c_{321}, c_{333}$$

and all of the remainders are false.

Don't be repelled by the number of atoms in the above example nor by the labour needed to write down the wff A. The point is that the problem *can* be faithfully described by a wff in PL, and the solution to the problem is found via a satisfying assignment of the atoms. Thus although PL is quite an impoverished language, it can nevertheless be used to describe quite complex problems. The 'trick' lies in the choice of the atoms. In many ways, PL is like an *assembly language* and it is perfectly adapted to studying a particular class of problems that are widespread and important, which we shall discuss in more detail in Section 1.9.

There is a final point. People solve Sudoku puzzles rather quickly (and even do so for entertainment) and are clearly not solving them using truth tables but are instead applying logical rules. We shall say more about this approach to logic in Section 3.3.

1.5.2 PL can be used to model some computers

In this section, we shall show that PL can be used to describe what are termed finite state automata. These can be viewed as the simplest mathematical models of computers.[15] We shall meet them again in Chapter 2, where they are used to describe the behaviour of computer circuits involving memory. My aim here is to convey the idea of such simple machines and show that their behaviour may be modelled using PL. The following motivating example is plagiarized from my book [43].

Given two coins, there are four possible ways of arranging them depending on which is heads (H) and which is tails (T):

$$HH, TH, HT, TT.$$

We call these the *states* of our system. Now consider the following two operations: 'flip the first coin', which I shall denote by a, and 'flip the second coin', which I shall denote by b. Assume that initially the coins are laid out as HH. This is our *initial state*. I am interested in all the possible ways of applying the operations a and b so that the coins are laid out as TT which will be our *terminal state*. The following diagram illustrates the relationships between the states and the two operations:

---

[15]They are, as it happens, special kinds of Turing machines which are the more refined mathematical models of computers. In any event, they are another legacy of Turing's work.

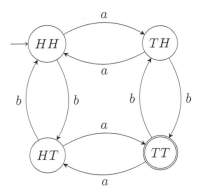

I have marked the start state with an inward-pointing arrow, and the terminal state by a double circle. If we start in the state HH and input the string *aba*, then we pass through the following states:

$$\text{HH} \xrightarrow{a} \text{TH} \xrightarrow{b} \text{TT} \xrightarrow{a} \text{HT}.$$

Thus the overall effect of starting at HH and inputting the string *aba* is to end up in the state HT. We say that a string over the alphabet $\{a, b\}$ is *accepted* if it labels a path from the initial state to a terminal state. It should be clear that in this case a string is accepted if and only if the number of *a*'s is odd and the number of *b*'s is odd. This is the *language accepted by this machine.*

If H is interpreted as 1 and T as 0 then our four states can be regarded as all the possible contents of two memory locations in a computer: 00, 10, 01, 11. The input *a* flips the input in the first location and the input *b* flips the input in the second location.

More generally, a finite state automaton has a finite number of states, represented in a diagram by circles that are usually labelled in some way, in which one state is singled out as the initial state and one or more states are identified as terminal states. Associated with the automaton is a finite alphabet $A = \{a_1, \ldots, a_n\}$. For each state s and each input letter a_i from A there is a uniquely determined next state t so that $s \xrightarrow{a_i} t$. We shall now describe how the behaviour of a finite state automaton can be encoded using PL.

Here is our original automaton but I have now numbered the states. The point is, we don't need to know the internal workings of the states — that they are represented by two coins — since the diagram tells us exactly how the automaton behaves.

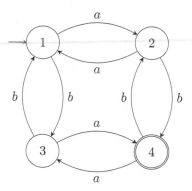

Our automaton changes state and so there is an implicit assumption that time will play a role. For us, time will be discretized and so will take values $t = 0, 1, 2, 3, \ldots$. At each time t, the automaton will be in exactly one of four states. Denote the state of the automaton at time t by $q(t)$. We can therefore write

$$\text{xor}\,((q(t) = 1), (q(t) = 2), (q(t) = 3), (q(t) = 4))$$

which does indeed say that at time t the automaton is in exactly one of its four states. If the automaton is in state $q(t)$ then its next state will depend on the input at time t. We denote the input at time t by $i(t)$. We can write

$$(i(t) = a) \oplus (i(t) = b)$$

because at time t either an a is input or a b is input, but not both.

> *It's important to observe that there are potentially infinitely many atoms (because the clock ticks: 0,1,2,3,...), but in processing a given input string we shall use only finitely many of them.*

We now have to describe the behaviour of the automaton in terms of PL. This is given by the following list of wff:

- $1 \xrightarrow{a} 2$ is described by the following wff

$$(q(t) = 1) \land (i(t) = a) \to (q(t+1) = 2).$$

- $1 \xrightarrow{b} 3$ is described by the following wff

$$(q(t) = 1) \land (i(t) = b) \to (q(t+1) = 3).$$

- $2 \xrightarrow{a} 1$ is described by the following wff

$$(q(t) = 2) \land (i(t) = a) \to (q(t+1) = 1).$$

- $2 \xrightarrow{b} 4$ is described by the following wff

$$(q(t) = 2) \land (i(t) = b) \to (q(t+1) = 4).$$

- $3 \xrightarrow{a} 4$ is described by the following wff

$$(q(t) = 3) \wedge (i(t) = a) \rightarrow (q(t+1) = 4).$$

- $3 \xrightarrow{b} 1$ is described by the following wff

$$(q(t) = 3) \wedge (i(t) = b) \rightarrow (q(t+1) = 1).$$

- $4 \xrightarrow{a} 3$ is described by the following wff

$$(q(t) = 4) \wedge (i(t) = a) \rightarrow (q(t+1) = 3).$$

- $4 \xrightarrow{b} 2$ is described by the following wff

$$(q(t) = 4) \wedge (i(t) = b) \rightarrow (q(t+1) = 2).$$

Finally, at $t = 0$ we start the automaton in its initial state. Thus $q(0) = 1$ is true.

Suppose we input a string of length 2. To do this, we need to specify $i(0)$ and $i(1)$. For example, if we input the string *ab* then the following wff is true: $(i(0) = a) \wedge (i(1) = b)$. The behaviour of our automaton in processing this string is now completely described by the conjunction of the following wff which we call A:

- Initial state: $q(0) = 1$.

- Input constraints:

$$((i(0) = a) \oplus (i(0) = b)) \wedge ((i(1) = a) \oplus (i(1) = b)).$$

- Unique state at each instant of time $t = 0$, $t = 1$ and $t = 2$:

$$\begin{aligned}
&\text{xor} \, ((q(0) = 1), (q(0) = 2), (q(0) = 3), (q(0) = 4)) \\
\wedge \quad &\text{xor} \, ((q(1) = 1), (q(1) = 2), (q(1) = 3), (q(1) = 4)) \\
\wedge \quad &\text{xor} \, ((q(2) = 1), (q(2) = 2), (q(2) = 3), (q(2) = 4)) \, .
\end{aligned}$$

- First set of possible state transitions:

$$\begin{aligned}
&(q(0) = 1) \wedge (i(0) = a) \rightarrow (q(1) = 2) \\
\wedge \quad &(q(0) = 1) \wedge (i(0) = b) \rightarrow (q(1) = 3) \\
\wedge \quad &(q(0) = 2) \wedge (i(0) = a) \rightarrow (q(1) = 1) \\
\wedge \quad &(q(0) = 2) \wedge (i(0) = b) \rightarrow (q(1) = 4) \\
\wedge \quad &(q(0) = 3) \wedge (i(0) = a) \rightarrow (q(1) = 4) \\
\wedge \quad &(q(0) = 3) \wedge (i(0) = b) \rightarrow (q(1) = 1) \\
\wedge \quad &(q(0) = 4) \wedge (i(0) = a) \rightarrow (q(1) = 3) \\
\wedge \quad &(q(0) = 4) \wedge (i(0) = b) \rightarrow (q(1) = 2).
\end{aligned}$$

- Second set of possible state transitions:

$$(q(1) = 1) \wedge (i(1) = a) \rightarrow (q(2) = 2)$$
$$\wedge \quad (q(1) = 1) \wedge (i(1) = b) \rightarrow (q(2) = 3)$$
$$\wedge \quad (q(1) = 2) \wedge (i(1) = a) \rightarrow (q(2) = 1)$$
$$\wedge \quad (q(1) = 2) \wedge (i(1) = b) \rightarrow (q(2) = 4)$$
$$\wedge \quad (q(1) = 3) \wedge (i(1) = a) \rightarrow (q(2) = 4)$$
$$\wedge \quad (q(1) = 3) \wedge (i(1) = b) \rightarrow (q(2) = 1)$$
$$\wedge \quad (q(1) = 4) \wedge (i(1) = a) \rightarrow (q(2) = 3)$$
$$\wedge \quad (q(1) = 4) \wedge (i(1) = b) \rightarrow (q(2) = 2).$$

- All the wff above are generic and can easily be written down for any string of length n where, here, $n = 2$. Finally, there is the actual input string of length 2:

$$(i(0) = a) \wedge (i(1) = b).$$

There is exactly one truth assignment making A true. This assigns the truth value T to the atoms $q(0) = 1$, $q(1) = 2$ and $q(2) = 4$ with all other truth assignments to the wff of the form $q(t) = k$ being false. Observe that a string of length n is accepted if and only if the state of the automaton at time $t = n$ is one of the terminal states. This, too, can easily be encoded by means of a wff in PL.

There are different kinds of automata. The automata defined in this section are called *finite state acceptors*. I shall use the term *finite state automaton* to mean simply an acceptor with initial and terminal states not marked. There are two kinds of ways in which outputs may be produced and these lead to two kinds of automata. A *Mealy machine* is a finite state automaton with an initial state marked and each transition produces an output. A *Moore machine* is a finite state automaton with an initial state marked and each state produces an output.

Exercises 1.5

---

1. The goal of this question is to construct a wff A for solving the following Sudoku-type problem. We use a 2×2 array, labelled as follows:

c_{11}	c_{12}
c_{21}	c_{22}

The atoms are:

- $p =$ 'c_{11} contains 1'.
- $q =$ 'c_{11} contains 2'.
- $r =$ 'c_{12} contains 1'.
- $s =$ 'c_{12} contains 2'.
- $t =$ 'c_{21} contains 1'.
- $u =$ 'c_{21} contains 2'.
- $v =$ 'c_{22} contains 1'.
- $w =$ 'c_{22} contains 2'.

(a) Construct A.

(b) How many rows will the truth table for A contain?

(c) For how many of these rows will the value of A be T?

(d) How will these rows relate to the solution of the Sudoku?

(e) Use the truth table generator [76] for your wff and check whether your answers to the above questions were correct.

2. This question will talk you through the PL solution to full Sudoku. Each Sudoku is a 9×9 grid of *cells*. Each cell is located by giving its *row number i* and its *column number j*. Thus we may refer to the cell c_{ij} where $1 \leq i \leq 9$ and $1 \leq j \leq 9$. Each cell can contain a digit k where $1 \leq k \leq 9$. There will therefore be $9 \times 9 \times 9 = 729$ atoms defined as follows:

$$p_{i,j,k} = \text{'}c_{ij} \text{ contains the digit } k\text{'}.$$

(a) To initialize the problem, define I to be a wff in the above atoms that describes the initial distribution of digits in the cells. Write down the wff I in the case of the Sudoku puzzle in Question 5 of the Introductory exercises. How many atoms are needed?

(b) Write down the wff $E_{i,j}$ which says that c_{ij} contains exactly one digit.

(c) Write down the wff E which says that every cell contains exactly one digit.

(d) Write down the wff R_i which says that row i contains each of the digits exactly once.

(e) Write down the wff R which says that in each row, each digit occurs exactly once in that row.

(f) Write down the wff C which says that in each column, each digit occurs exactly once in that column.

(g) The Sudoku grid is also divided into 9 *blocks* each containing 3×3 cells. Number these blocks $1, 2, \ldots, 9$ from top to bottom and left to right. Write down the wff W_l which says that in block l each of the nine digits occurs exactly once.

(h) Write down the wff W which says that in each block, each of the digits occurs exactly once.

(i) Put $P = I \wedge E \wedge R \wedge C \wedge W$. Determine under what circumstances the wff P is satisfiable.

3. Write down all the wff that describe the following finite state acceptor:

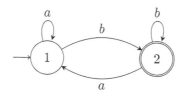

Show how your wff lead to the correct processing of the input string ab.

1.6 Adequate sets of connectives

We defined our version of PL using the following six connectives

$$\neg, \wedge, \vee, \rightarrow, \leftrightarrow, \oplus.$$

This is not a unique choice: for example, many books on logic do not include $\oplus$. In this section, we shall explore what is actually needed to define PL.

In Examples 1.4.4, we showed that

$$p \oplus q \equiv \neg(p \leftrightarrow q)$$

and

$$p \leftrightarrow q \equiv (p \rightarrow q) \wedge (q \rightarrow p)$$

as well as

$$p \rightarrow q \equiv \neg p \vee q.$$

We have therefore proved the following.

Proposition 1.6.1. *Every wff in PL is logically equivalent to one that uses only the connectives $\neg$, $\vee$ and $\wedge$.*

Although it would be more efficient to use only the above three connectives, it would also be less user-friendly. The important point is, however, that those books that do not take $\oplus$ as a basic connective are not sacrificing any expressive power.

At this point, we introduce some terminology. We say that a set of connectives is *adequate* if every wff is logically equivalent to a wff that uses only the connectives from that set. In these terms, we proved above that the connectives $\neg, \vee, \wedge$ form an adequate set.

We can, if we want, be even more parsimonious in the number of connectives we use. The following two logical equivalences can be proved using double negation and De Morgan.

- $p \vee q \equiv \neg(\neg p \wedge \neg q)$.

- $p \wedge q \equiv \neg(\neg p \vee \neg q)$.

From these we can deduce the following.

Proposition 1.6.2.

1. *The connectives $\neg$ and $\wedge$ together form an adequate set.*

2. *The connectives $\neg$ and $\vee$ together form an adequate set.*

Examples 1.6.3. The following wff are equivalent to wff using only the connectives $\neg$ and $\wedge$.

1. $p \vee q \equiv \neg(\neg p \wedge \neg q)$.

2. $p \to q \equiv \neg p \vee \neg\neg q \equiv \neg(p \wedge \neg q)$.

3. $p \leftrightarrow q \equiv (p \to q) \wedge (q \to p) \equiv \neg(p \wedge \neg q) \wedge \neg(q \wedge \neg p)$.

At this point, you might wonder if we can go one better. Indeed we can but we have to introduce two new binary connectives. Define

$$p \downarrow q = p \, \mathbf{nor} \, q = \neg(p \vee q)$$

called *nor*. Define

$$p \mid q = p \, \mathbf{nand} \, q = \neg(p \wedge q)$$

called *nand*.

Proposition 1.6.4.

1. *The binary connective $\downarrow$ is adequate on its own.*

2. *The binary connective $\mid$ is adequate on its own.*

Proof. (1) Observe that

$$\neg p \equiv \neg p \wedge \neg p \equiv \neg(p \vee p) \equiv p \downarrow p$$

and

$$p \wedge q \equiv \neg\neg p \wedge \neg\neg q \equiv \neg(\neg p \vee \neg q) \equiv (\neg p) \downarrow (\neg q) \equiv (p \downarrow p) \downarrow (q \downarrow q).$$

But since $\neg$ and $\wedge$ form an adequate set of connectives we can now construct everything using $\downarrow$ alone.

(2) Observe that

$$\neg p \equiv \neg p \vee \neg p \equiv \neg(p \wedge p) \equiv p \mid p$$

and

$$p \vee q \equiv \neg\neg p \vee \neg\neg q \equiv \neg(\neg p \wedge \neg q) \equiv (\neg p) \mid (\neg q) \equiv (p \mid p) \mid (q \mid q). \qquad \square$$

It would be possible to develop the whole of PL using, for example, only **nor**, and some mathematicians have done just that, but this makes PL truly non-user-friendly.

Example 1.6.5. We find a wff logically equivalent to $p \to q$ that uses only **nor**s.

$$\begin{aligned} p \to q & \equiv & \neg p \vee q \\ & \equiv & \neg\neg(\neg p \vee q) \\ & \equiv & \neg(\neg(\neg p \vee q)) \\ & \equiv & \neg(\neg p \downarrow q) \\ & \equiv & \neg((p \downarrow p) \downarrow q) \\ & \equiv & ((p \downarrow p) \downarrow q) \downarrow ((p \downarrow p) \downarrow q). \end{aligned}$$

You should supply reasons that justify each line above.

Exercises 1.6

---

1. Write each of the connectives $\neg, \wedge, \vee, \to$ in terms of **nand** only. Now, repeat this exercise using **nor** only.

2. (a) Show that $\neg$ and $\to$ together form an adequate set of connectives.
   (b) Show that $\to$ and $\boldsymbol{f}$ together form an adequate set of connectives.

3. Let $*$ be a binary connective that is supposed to be adequate on its own. There are 16 possible truth tables for $*$.

   (a) Explain why $T * T = F$.
   (b) Explain why $F * F = T$.
   (c) There are now four possible ways to complete the truth table for $*$. By examining what they are, deduce that the only possible values for $*$ are $\downarrow$ and $\mid$.

1.7 Truth functions

This section is a companion piece to the preceding one. We now have the ideas required to prove a result significant in circuit design. It will also tell us that PL is 'functionally complete', a phrase explained later in this section. But before I can do this, I need to introduce a couple of notions from mathematics.

Functions

Let A and B be sets. A *function* from A to B is a rule f that determines for each element of A a uniquely specified element of B. We often write $f \colon A \to B$ as a way of expressing the fact that f is a function from A to B where the set A, called the *domain*, is the set of allowable inputs, and the set B, called the *codomain*, is the set that contains all the outputs (and possibly more besides). If $a \in A$ then the *value* of f at a is written $f(a)$ and the function can also be written as $a \mapsto f(a)$ which should be read as 'a maps to $f(a)$'.

Functions are the bread and butter of mathematics and the dollars and cents of computer science — it is sometimes useful to think of a programme as computing a function that transforms inputs to outputs. However, in this section we shall only consider a very simple class of functions and to define those we need some further definitions.

Products of sets

In Section 1.2, I defined sets. You will recall that in set notation $\{a, b\} = \{b, a\}$, since the order in which we list the elements of the set doesn't matter, but there are lots of situations where order *does* matter. For example, in a race we want to know who comes first, who comes second and who comes third, not merely that three named people crossed the winning tape. The idea needed to express this is different. An *ordered pair* (a, b) is a pair of elements a and b where a is the *first element* and b is the *second element*. Thus, in general, $(a, b) \neq (b, a)$ unless $a = b$, of course. More generally, we can define *ordered triples* (a, b, c) and *ordered 4-tuples* (a, b, c, d) and, more generally, *ordered n-tuples* $(a_1, \ldots, a_n)$.

I should add that ordered n-tuples are also called *lists* of length n. In addition, it will occasionally be more convenient to use strings of length n to represent ordered n-tuples.

We can use these ideas to define a way of combining sets. Let A and B be sets. Define the set

$$A \times B$$

to be the set of all ordered pairs (a, b) where $a \in A$ and $b \in B$. This is called the *product of A and B*. More generally, if $A_1, \ldots, A_n$ are sets, define the set

$$A_1 \times \ldots \times A_n$$

to be the set of all ordered n-tuples $(a_1, \ldots, a_n)$ where $a_1 \in A_1$, and $a_2 \in A_2$ and, $\ldots$, $a_n \in A_n$. If $A = A_1 = \ldots = A_n$ then we write A^n for the set of all n-tuples whose elements are taken from A.

Example 1.7.1. Let $A = \{1, 2, 3\}$ and let $B = \{a, b\}$. Then the product set, $A \times B$, is given by

$$A \times B = \{(1, a), (1, b), (2, a), (2, b), (3, a), (3, b)\}.$$

On the other hand, the product set $B \times A$ is given by

$$B \times A = \{(a, 1), (b, 1), (a, 2), (b, 2), (a, 3), (b, 3)\}.$$

Thus $A \times B \neq B \times A$.

Now that we have defined functions and products of sets, we can make the key definition of this section. Put $\mathbb{T} = \{T, F\}$. *Any function $f \colon \mathbb{T}^n \to \mathbb{T}$ is called a truth function.* Because the domain of this function is $\mathbb{T}^n$ we say that it has n *arguments*; this simply means that the function has n inputs. Although I have used mathematical terms to define truth functions, what they are in practice is easy to explain. The set $\mathbb{T}^n$ is essentially the set of all possible lists of n truth values. There are 2^n of those by Lemma 1.3.3. For each such list of ordered truth values, the function f assigns a single truth value. Thus each truth function determines a truth table and, conversely, each truth table determines a truth function. However, when I say 'truth table' I am not assuming that there is a wff that has that truth table. But the essence of the main theorem of this section is, that in fact, there is.

Example 1.7.2. Here is an example of a truth function $f \colon \mathbb{T}^3 \to \mathbb{T}$.

T	T	T	T
T	T	F	F
T	F	T	F
T	F	F	T
F	T	T	T
F	T	F	F
F	F	T	F
F	F	F	F

For example, $f(T, F, F) = T$ and $f(F, F, F) = F$.

Theorem 1.7.3 (Functional completeness). *For each truth function with n arguments there is a wff with n atoms whose truth table is the given truth function.*

Proof. We shall prove this result in three steps constructing a wff A.

Step 1. Suppose that the truth function always outputs F. Define

$$A = (p_1 \land \neg p_1) \land p_2 \land \ldots \land p_n.$$

Then A has a truth table with 2^n rows that always outputs F.

Step 2. Suppose that the truth function outputs T *exactly once*. Let $v_1, \ldots, v_n$, where $v_i = T$ or F, be the assignment of truth values which yields the output T. Define a wff A as follows. It is a conjunction of exactly one of p_1 or $\neg p_1$, of p_2 or $\neg p_2, \ldots$, of p_n or $\neg p_n$ where p_i is chosen if $v_i = T$ and $\neg p_i$ is chosen if $v_i = F$. I shall call A a *conjunctive clause* corresponding to the pattern of truth values $v_1, \ldots, v_n$. The truth table of A is the given truth function.

Step 3. Suppose that we are given now an arbitrary truth function not covered in steps 1 and 2 above. We construct a wff A whose truth table is the given truth function by taking a disjunction of all the conjunctive clauses constructed from each row of the truth function that outputs T. That this construction does what is required follows from the observation that a conjunctive clause is true for exactly one truth assignment to its atoms. $\square$

The proof of the above theorem is best explained by means of an example.

Example 1.7.4. We construct a wff that has as truth table the following truth function:

T	T	T	$\boldsymbol{T}\,(1)$
T	T	F	F
T	F	T	F
T	F	F	$\boldsymbol{T}\,(2)$
F	T	T	$\boldsymbol{T}\,(3)$
F	T	F	F
F	F	T	F
F	F	F	F

We need only consider the rows that output T, which I have highlighted. I have also included a reference number that I shall use below. The conjunctive clause corresponding to row (1) is

$$p \land q \land r.$$

The conjunctive clause corresponding to row (2) is

$$p \land \neg q \land \neg r.$$

The conjunctive clause corresponding to row (3) is

$$\neg p \land q \land r.$$

The disjunction of these conjunctive clauses is

$$A = (p \wedge q \wedge r) \vee (p \wedge \neg q \wedge \neg r) \vee (\neg p \wedge q \wedge r).$$

You should check that the truth table of A is the truth function given above.

Exercises 1.7

---

1. The following truth table describes the behaviour of three truth functions (a), (b) and (c). In each case, find a wff whose truth table is equal to the corresponding truth function.

p	q	r	(a)	(b)	(c)
T	T	T	F	F	T
T	T	F	F	T	F
T	F	T	T	F	T
T	F	F	F	T	F
F	T	T	F	F	T
F	T	F	F	F	F
F	F	T	F	F	T
F	F	F	F	F	F

2. Prove that $\{\{a\}, \{a, b\}\} = \{\{c\}, \{c, d\}\}$ if and only if $a = c$ and $b = d$. Deduce that we may *define* $(a, b) = \{\{a\}, \{a, b\}\}$. *I like this result because it shows that order may be defined in terms of un-order.*

3. Show that ordered n-tuples, where $n \geq 3$, may be defined in terms of ordered pairs.

1.8   Normal forms

I could, if I were feeling mischievous, write $\neg\neg\neg\neg\neg p$ instead of $\neg p$. Writing the former rather than the latter would not be normal, but neither would it be disallowed. However, good communication depends on agreements on how we do things. This section is about particular ways of writing wff called *normal forms*.

1.8.1   Negation normal form (NNF)

A wff is in *negation normal form (NNF)* if it is constructed using only $\wedge$, $\vee$ and literals. Recall that a *literal* is either an atom or the negation of an atom.

Proposition 1.8.1. *Every wff is logically equivalent to a wff in NNF.*

Proof. Let A be a wff in PL. First, replace all occurrences of $x \oplus y$ by $\neg(x \leftrightarrow y)$. Second, replace all occurrences of $x \leftrightarrow y$ by $(x \to y) \wedge (y \to x)$. Third, replace all occurrences of $x \to y$ by $\neg x \vee y$. Fourth, use De Morgan's laws to push all occurrences of negation through brackets. Finally, use double negation to ensure that only literals occur. $\square$

Example 1.8.2. We convert $\neg(p \to (p \wedge q))$ into NNF using the method outlined in the above proof.

$$
\begin{aligned}
\neg(p \to (p \wedge q)) &\equiv \neg(\neg p \vee (p \wedge q)) \\
&\equiv \neg\neg p \wedge \neg(p \wedge q) \\
&\equiv \neg\neg p \wedge (\neg p \vee \neg q) \\
&\equiv p \wedge (\neg p \vee \neg q).
\end{aligned}
$$

1.8.2 Disjunctive normal form (DNF)

We now come to the first of the two important normal forms. A wff that can be written as a disjunction of one or more terms, each of which is a conjunction of one or more literals, is said to be in *disjunctive normal form (DNF)*. Thus a wff in DNF has the following schematic shape:

$$(\wedge \text{ literals}) \vee \ldots \vee (\wedge \text{ literals}).$$

Examples 1.8.3. Some special cases of DNF are worth highlighting because they often cause confusion.

1. A single atom p is in DNF.

2. A term such as $(p \wedge q \wedge \neg r)$ is in DNF.

3. The expression $p \vee q$ is in DNF. You should think of it as $(p) \vee (q)$.

Proposition 1.8.4. *Every wff is logically equivalent to one in DNF.*

Proof. Let A be a wff. Construct the truth table for A. Now apply Theorem 1.7.3. The wff that results is in DNF and logically equivalent to A. $\square$

The proof above can be also be used as a method for constructing DNF, though it is a little laborious. Another method is to use logical equivalences, as follows. Let A be a wff. First convert A to NNF and then if necessary use the distributive laws to convert to a wff which is in DNF.

Example 1.8.5. We show how to convert $\neg(p \to (p \wedge q))$ into DNF using a sequence of logical equivalences. By Example 1.8.2, we know that $p \wedge (\neg p \vee \neg q)$ is the NNF of this wff. We now apply one of the distributive laws to get the $\vee$ out of the brackets. This yields $(p \wedge \neg p) \vee (p \wedge \neg q)$. This wff is in DNF and so

$$\neg(p \to (p \wedge q)) \equiv (p \wedge \neg p) \vee (p \wedge \neg q).$$

1.8.3 Conjunctive normal form (CNF)

A wff is in *conjunctive normal form (CNF)* if it is a conjunction of one or more terms, each of which is a disjunction of one or more literals. It therefore looks a bit like the reverse of DNF:

$$(\vee \text{ literals}) \wedge \ldots \wedge (\vee \text{ literals}).$$

Proposition 1.8.6. *Every wff is logically equivalent to one in CNF.*

Proof. Let A be our wff. Write $\neg A$ in DNF by Proposition 1.8.4. Now negate both sides of the logical equivalence and use double negation where necessary to obtain a wff in CNF. □

Example 1.8.7. A wff A has the following truth table:

p	q	r	A
T	T	T	T
T	T	F	T
T	F	T	F
T	F	F	T
F	T	T	T
F	T	F	F
F	F	T	T
F	F	F	T

The truth table for $\neg A$ is

p	q	r	$\neg A$
T	T	T	F
T	T	F	F
T	F	T	T
T	F	F	F
F	T	T	F
F	T	F	T
F	F	T	F
F	F	F	F

It follows that

$$\neg A \equiv (p \wedge \neg q \wedge r) \vee (\neg p \wedge q \wedge \neg r)$$

is the DNF for $\neg A$. Negating both sides we get

$$A \equiv (\neg p \vee q \vee \neg r) \wedge (p \vee \neg q \vee r).$$

This is the CNF for A.

Sometimes using logical equivalences is more efficient.

Example 1.8.8. The wff $(\neg x \wedge \neg y) \vee (\neg x \wedge z)$ is in DNF. It can easily be converted into CNF using one of the distributivity laws.

$$
\begin{aligned}
(\neg x \wedge \neg y) \vee (\neg x \wedge z) &\equiv ((\neg x \wedge \neg y) \vee \neg x) \wedge ((\neg x \wedge \neg y) \vee z) \\
&\equiv (\neg x \vee \neg x) \wedge (\neg y \vee \neg x) \wedge (\neg x \vee \neg z) \wedge (\neg y \wedge z).
\end{aligned}
$$

1.8.4 Prologue to PROLOG

Computer programs are written using artificial languages called *programming languages*. There are two big classes of such languages: *imperative* and *declarative*.

- In an imperative language, the solution to a problem has to be spelt out in detail. In other words, the computer has to be told *how* to solve the problem.

- In a declarative language, the computer is told *what* to solve and the details of how to accomplish this are left to the software.

Declarative languages sound like magic but some of them, such as PROLOG, are based on ideas drawn from first-order logic. This language has applications in AI (artificial intelligence). We describe here the wff in PL that are important in PROLOG.

A literal is *positive* if it is an atom and *negative* if it is the negation of an atom. A wff in CNF is called a *Horn formula*[16] if each group of disjunctions contains at most one positive literal. This seems a little arbitrary but becomes more natural when implications and the logical constants are used. Here are some examples.

1. $\neg p \vee q \equiv p \to q$.

2. $\neg p \vee \neg q \vee r \equiv p \wedge q \to r$.

3. The wff $\neg p \vee \neg q$ is at first glance more puzzling but if we use the logical constant $\boldsymbol{f}$, we have the following

$$\neg p \vee \neg q \equiv \neg p \vee \neg q \vee \boldsymbol{f} \equiv p \wedge q \to \boldsymbol{f}.$$

4. Likewise, using the logical constant $\boldsymbol{t}$ we have the following sequence of logical equivalences

$$p \equiv \boldsymbol{f} \vee p \equiv \neg \boldsymbol{t} \vee p \equiv \boldsymbol{t} \to p.$$

The above examples illustrate that every Horn formula is logically equivalent to a conjunction of wff, each of which has one of the following three forms:

---

[16]Named after the American mathematician Alfred Horn (1918–2001).

- *Rules:* $p_1 \wedge \ldots \wedge p_n \rightarrow p$.

- *Goals:* $p_1 \wedge \ldots \wedge p_n \rightarrow \boldsymbol{f}$.

- *Facts:* $\boldsymbol{t} \rightarrow p$.

We call this the *implicational form* of the Horn formula.

Example 1.8.9. Applications of Horn formulae range from diagnosing ailments in people to fault location in machines (such as the reasons why your car might not start in the morning). As a concrete example, consider the following rules for animal taxonomy:

- furry $\wedge$ animal $\rightarrow$ mammal.

- scaly $\wedge$ animal $\rightarrow$ reptile.

- mammal $\wedge$ eats carrots $\rightarrow$ rabbit.

- reptile $\wedge$ breathes fire $\rightarrow$ dragon.

You then spot an unknown animal which you observe has the following characteristics: $\boldsymbol{t} \rightarrow$ furry, $\boldsymbol{t} \rightarrow$ animal, and $\boldsymbol{t} \rightarrow$ eats carrots. Then you are able to deduce that the animal is a rabbit rather than, say, a dragon.

Example 1.8.10. Consider the wff

$$H = (p \vee \neg s \vee \neg u) \wedge (q \vee \neg r) \wedge (\neg q \vee \neg s) \wedge (s) \wedge (\neg s \vee u).$$

This is a Horn formula and it can be written in its implicational form as follows:

$$H = (s \wedge u \rightarrow p) \wedge (r \rightarrow q) \wedge (q \wedge s \rightarrow \boldsymbol{f}) \wedge (\boldsymbol{t} \rightarrow s) \wedge (s \rightarrow u).$$

One of the reasons Horn formulae are interesting is that there is a fast algorithm for deciding whether they are satisfiable or not; in the next section, we shall explain why this is unlikely to be true for arbitrary wff.

Fast algorithm for Horn formula satisfiability

Let the Horn formula be $H = H_1 \wedge \ldots \wedge H_m$ where each H_i is either a rule, a goal or a fact. Let the atoms occurring in H be $p_1, \ldots, p_n$. The algorithm begins by initially assigning all the atoms the truth value F. If the algorithm terminates successfully, then some of these truth values might have changed to T, but in any event the final truth values will satisfy H. During the algorithm we shall *mark* atoms which can be done by placing a dot above them, like so $\dot{p}$. The algorithm is as follows:

1. For each H_i of the form $\boldsymbol{t} \rightarrow p$, mark all occurrences of the atom p in H.

2. The following procedure is now repeated until it can no longer be applied:

   - If there is an H_i of the form $p_1 \wedge \ldots \wedge p_n \rightarrow p$ where all of $p_1, \ldots, p_n$ have been marked and p has not been marked, then mark every occurrence of p.
   - If there is an H_i of the form $p_1 \wedge \ldots \wedge p_n \rightarrow \boldsymbol{f}$ where all of $p_1, \ldots, p_n$ have been marked, then terminate the algorithm and output that H is *not satisfiable*.

3. Output that H *is satisfiable*; all marked atoms should be assigned the truth value T (with the remaining atoms keeping their initially assigned truth value of F), and this results in a truth assignment that satisfies H.

Before we prove that the above algorithm really does what I claim it does, we give an example of it in operation.

Example 1.8.11. We apply the above algorithm to the Horn formula

$$H = (s \wedge u \rightarrow p) \wedge (r \rightarrow q) \wedge (q \wedge s \rightarrow \boldsymbol{f}) \wedge (t \rightarrow s) \wedge (s \rightarrow u).$$

We apply part (1) of the algorithm and so mark all occurrences of the atom s:

$$H = (\dot{s} \wedge u \rightarrow p) \wedge (r \rightarrow q) \wedge (q \wedge \dot{s} \rightarrow \boldsymbol{f}) \wedge (t \rightarrow \dot{s}) \wedge (\dot{s} \rightarrow u).$$

We now apply part (2) of the algorithm repeatedly. First, we mark all occurrences of the atom u:

$$H = (\dot{s} \wedge \dot{u} \rightarrow p) \wedge (r \rightarrow q) \wedge (q \wedge \dot{s} \rightarrow \boldsymbol{f}) \wedge (t \rightarrow \dot{s}) \wedge (\dot{s} \rightarrow \dot{u}).$$

Second, we mark all occurrences of the atom p:

$$H = (\dot{s} \wedge \dot{u} \rightarrow \dot{p}) \wedge (r \rightarrow q) \wedge (q \wedge \dot{s} \rightarrow \boldsymbol{f}) \wedge (t \rightarrow \dot{s}) \wedge (\dot{s} \rightarrow \dot{u}).$$

At this point, the algorithm terminates successfully. A satisfying truth assignment is therefore

p	q	r	s	u
T	F	F	T	T

If you run the truth table generator [76] on this example, you will find that this truth assignment is the first one that satisfies H starting from the bottom of the truth table, although in this case it is in fact the only satisfying truth assignment.

Example 1.8.12. We apply the above algorithm to the following Horn formula

$$H = (p \wedge q \to \boldsymbol{f}) \wedge (t \to p) \wedge (t \to q).$$

This yields

$$(\dot{p} \wedge \dot{q} \to \boldsymbol{f}) \wedge (t \to \dot{p}) \wedge (t \to \dot{q})$$

which immediately tells us that H is a contradiction because of the first term of H.

Example 1.8.13. We apply the above algorithm to the following Horn formula

$$H = (p \wedge q \wedge s \to \boldsymbol{f}) \wedge (q \wedge r \to \boldsymbol{f}) \wedge (s \to \boldsymbol{f}).$$

The algorithm terminates as soon as it has started since there are no atoms to be marked. It follows that a satisfying truth assignment is one that assigns the truth value F to all atoms.

In the proof that follows, it is important to remember that $x \to y$ is true except when x is true and y is false.

Theorem 1.8.14. *The algorithm above, for deciding whether a Horn formula is satisfiable, works.*

Proof. Let H be a Horn formula. Observe first that if H does not contain any wff of the form $t \to p$, then the algorithm terminates successfully immediately after it starts with the wff being satisfied by the truth assignment where every atom is F. We shall therefore assume in what follows that H contains at least one wff of the form $t \to p$. Call an atom *good* if it is eventually marked after repeated applications of the following two procedures:

1. Mark all atoms p where $t \to p$ occurs in H.

2. Mark all atoms q whenever $p_1 \wedge \ldots \wedge p_n \to q$ occurs in H as one of the H_j and all the atoms $p_1, \ldots, p_n$ have been marked.

We now derive an important property of Horn formulae: in any truth assignment τ satisfying H, the good atoms must be assigned the truth value T. To see why this property holds, observe from the definition of implication that the only way for τ to satisfy $t \to p$ is for τ to assign the truth value T to p; if

τ assigns the truth value T to $p_1, \ldots, p_n$, then it must assign the truth value T to q in order that it assign the truth value T to the wff $p_1 \wedge \ldots \wedge p_n \to q$. It follows that if τ_1 and τ_2 are truth assignments making H true, then they must both make all the good atoms true.

Let τ^* be the truth assignment making all the good atoms true (and all the remaining atoms false). This is the truth assignment constructed by the algorithm. There are now two possibilities:

- Suppose that H contains no wff of the form $p_1 \wedge \ldots \wedge p_n \to \boldsymbol{f}$. Then τ^* satisfies H.

- Suppose that H contains at least one wff of the form $p_1 \wedge \ldots \wedge p_n \to \boldsymbol{f}$. There are now two possibilities: if one of these wff is such that all the atoms $p_1, \ldots, p_n$ are good then H is not satisfiable otherwise τ^* satisfies H. $\qquad\square$

Exercises 1.8

---

1. Use known logical equivalences to transform each of the following wff first into NNF and then into DNF.

   (a) $(p \to q) \to p$.

   (b) $p \to (q \to p)$.

   (c) $(q \wedge \neg p) \to p$.

   (d) $(p \vee q) \wedge r$.

   (e) $p \to (q \wedge r)$.

   (f) $(p \vee q) \wedge (r \to s)$.

2. Use known logical equivalences to transform each of the following into CNF.

   (a) $(p \to q) \to p$.

   (b) $p \to (q \to p)$.

   (c) $(q \wedge \neg p) \to p$.

   (d) $(p \vee q) \wedge r$.

   (e) $p \to (q \wedge r)$.

   (f) $(p \vee q) \wedge (r \to s)$.

3. Let $A = ((p \wedge q) \to r) \wedge (\neg(p \wedge q) \to r)$.

   (a) Draw the truth table for A.
   (b) Construct DNF using (a).
   (c) Draw the truth table for $\neg A$.
   (d) Construct DNF for $\neg A$ using (c).
   (e) Construct CNF for A using (d).

4. Let $A = ((p \wedge q) \to r) \wedge (\neg(p \wedge q) \to r)$.

   (a) Write A in NNF.
   (b) Use known logical equivalences applied to (a) to get DNF.
   (c) Use logical equivalences applied to (a) to get CNF.
   (d) Simplify A as much as possible using known logical equivalences.

5. Write $p \leftrightarrow (q \leftrightarrow r)$ in NNF.

6. For each of the following Horn formulae, write them in implicational form and then use the algorithm described in the text to determine whether they are satisfiable or not. In each case, check your answer using the truth table generator [76].

   (a) $(p \vee \neg q) \wedge (q \vee \neg r)$.
   (b) $(p \vee \neg q \vee \neg r) \wedge (\neg s \vee \neg u) \wedge (\neg p \vee \neg q \vee r) \wedge (p) \wedge (q)$.
   (c) $(p) \wedge (q) \wedge (\neg p \vee \neg q) \wedge (r \vee \neg p)$.

7. Prove that $p \oplus q$ is not logically equivalent to any Horn formula. To answer this question, you should focus on the important property of Horn formulae described in the proof of Theorem 1.8.14.

1.9 $\mathcal{P} = \mathcal{NP}$? or How to win a million dollars

The question whether $\mathcal{P}$ is equal to $\mathcal{NP}$ is the first of the seven Millennium Prize Problems that were posed by the Clay Mathematics Institute[17] in the year 2000. Anyone who solves one of these problems wins a million dollars. So far, only one of these problems has been solved: namely, the *Poincaré Conjecture* by Grigori Perelman who turned down the prize money [21]. The other six problems require advanced mathematics just to understand what they are saying — except one. This is the question of whether $\mathcal{P}$ is equal to

---

[17]http://www.claymath.org/.

$\mathcal{NP}$. In this section, I shall describe this very question in intuitive terms and connect it with SAT.

We begin with a basic question in computer science: how can we measure how long it takes for a program to solve a problem? As it stands, this is too vague to admit an answer so we need to make it more precise. For concreteness, imagine a program that takes as input a whole number n and produces as output either the result that 'n is prime' or 'n is not prime'. So, if you input 10 to the program it would tell you it was not prime, but if you input 17 it would tell you that it was prime. Clearly, how long the program takes to solve this question depends on how big the number is that you input, since a number with hundreds of digits is clearly going to take a lot longer to be processed than one with just a few digits.

> Thus to say how long a program takes to solve a problem must refer to the length of the input to that program.

Now, for all inputs of a fixed length, say m, the program might take different amounts of time to produce an output depending on which input of length m was chosen.

> We agree to take the longest amount of time over all inputs of length m as a measure of how long it takes to process any input of length m.

This raises the question of what we mean by 'time' since your fancy MacBook Pro will be a lot faster than my Babbage Imperial. So instead of time, we count the number of basic computational steps needed to transform input to output. It turns out that we don't really have to worry too much about what this means, but it can be made precise using Turing machines.

Thus with each program we can try to calculate its *time complexity profile*. This will be a function $m \mapsto f(m)$ where m is the length of the input and $f(m)$ is the maximum number of steps needed to transform any input of size m into an output. Calculating the time complexity profile exactly of even simple programs requires a lot of mathematical analysis, but fortunately we don't need an exact answer, merely a good approximate one. The time complexity profile of a program is therefore any such good approximate one. A program that has the time complexity profile $m \mapsto am$, where a is a number, is said to run in *linear time*; if the time complexity profile is $m \mapsto am^2$, it is said to run in *quadratic time*; if the time complexity profile is $m \mapsto am^3$, it is said to run in *cubic time*. More generally, if the time complexity profile is $m \mapsto am^n$ for some n, it is said to run in *polynomial time*. All the basic algorithms you learnt at school for adding, subtracting, multiplying and dividing numbers run in polynomial time.

> We define the class $\mathcal{P}$ to be all those problems that can be solved in polynomial time.

58

These are essentially nice problems with nice (meaning fast) programs to solve them. Let me stress that just because you have an algorithm for a problem that is slower than polynomial time does not mean your problem does not have a polynomial time algorithm. If you claim this then you have to prove it.

Example 1.9.1. The problem I introduced above, of deciding whether a number is prime or not, does actually belong to the class $\mathcal{P}$ but this was only proved in 2004 [2]. The construction the authors develop is elementary but ingenious.

What would constitute a nasty problem? This would be one whose time complexity profile looked like $m \mapsto 2^m$, such as in Example 1.3.4. This is nasty because just by increasing the size of the input by 1 doubles the amount of time needed to solve the problem. We now isolate those nasty problems that have a nice feature: that is, any purported solution can be *checked* quickly, meaning that it can be checked in polynomial time. A really nasty problem would be one where it is even difficult just to check a purported solution.

> We define the class $\mathcal{NP}$, of *non-deterministic polynomial time problems*, to be all those problems whose solutions can be checked in polynomial time.

Clearly, $\mathcal{P}$ is contained in $\mathcal{NP}$ but there is no earthly reason why they should be equal. Now here is the rub: currently (2018) no one has been able to prove that they are not equal. This describes what, in essence, the question of whether $\mathcal{P}$ is equal to $\mathcal{NP}$ means. It does not, however, explain the significance of this question. To understand this, we must delve deeper.

In 1971, Stephen Cook[18] came up with an idea for resolving this question that shed new light on the nature of algorithms. His idea was to find a problem inside $\mathcal{NP}$ with the following property: if that problem could be shown to be in $\mathcal{P}$ then everything in $\mathcal{NP}$ would also have to be in $\mathcal{P}$ thus showing that $\mathcal{P} = \mathcal{NP}$. On the other hand, if it could be shown that this problem wasn't in $\mathcal{P}$ then we would, of course, have shown that $\mathcal{P} \neq \mathcal{NP}$. Thus Cook's problem would provide a sort of litmus test for determining whether $\mathcal{P} \neq \mathcal{NP}$. Such a problem is called $\mathcal{NP}$-*complete*. Given that we don't even know all the problems in $\mathcal{NP}$, Cook's idea sounds unworkable. But, in fact, he was able to prove a specific problem to be $\mathcal{NP}$-complete: that problem is SAT — the *satisfiability problem in PL*. This probably sounds mysterious but it essentially boils down to the fact that PL is a sufficiently rich language to describe the behaviour of computers — generalizing what we discussed in Section 1.5.2. Cook's result, known as *Cook's theorem*, is remarkable enough and explains the central importance of the satisfiability problem in theoretical computer science. But there is more, and it is this that really makes the question of

---

[18]And, independently at the same time in the Soviet Union, Leonid Levin. This sort of West/East mathematical pairing was extremely common during the Cold War. See [34].

whether $\mathcal{P} = \mathcal{NP}$ interesting: thousands of important, natural problems that we would like to solve quickly have been shown to be $\mathcal{NP}$-complete and so equivalent to SAT. Here are some examples.[19] The following definition will be useful. A *graph* consists of *vertices* which are represented by circles and *edges* which are lines joining the circles (here always joining different circles). Vertices joined by an edge are said to be *adjacent*. Here is an example of such a graph:

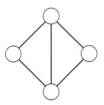

Examples 1.9.2. The following problems are all NP-complete.

1. *The travelling salesman problem.* A travelling salesman has to visit a number of cities starting and finishing at his home city. We call this the *route* the travelling salesman takes. More mathematically, we are given a graph G such that between any two vertices there is an edge labelled with a positive whole number which represents the distance between those two cities; we also allow the distance to be ∞, infinity, which is equivalent to saying that the two vertices are not really joined by an edge. In addition, we are given a positive whole number D. The question is: is there a route whose total distance is no greater than D?

2. *Quadratic Diophantine equations.* Given positive integers a, b, c are there positive integers x and y such that $ax^2 + by = c$?

3. *Latin squares.* A *Latin square* is an $n \times n$ grid of cells in which each cell contains a number between 1 and n such that in each row and each column, each number appears exactly once. A *partial Latin square* is an $n \times n$ grid of cells in which some of the cells contain a number between 1 and n such that in each row and each column, each number appears exactly once. The question is: can each partial Latin square be completed to a Latin square [9]?

4. *Sudoku.* Given an $n^2 \times n^2$ grid of cells divided into n blocks, where some of the cells contain numbers in the range of 1 to n^2, complete the grid in accordance with the usual Sudoku constraints. There is a close connection with the Latin square problem above [80].

5. *Minesweeper.* This is a familar game on the computer. I refer you to Richard Kaye's website for more information [39].

---

[19] The standard reference to all this is [19].

6. *Is a graph k-colourable?* By a *colouring* of a graph we mean an assignment of colours to the vertices so that adjacent vertices have different colours. By a *k-colouring* of a graph we mean a colouring that uses at most k colours. Depending on the graph and k that may or may not be possible. The following is a 3-colouring of the graph above.[20]

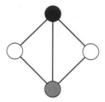

However, it is not possible to find a 2-colouring of this graph because there is a triangle of vertices. The crucial point is this: if you do claim to have found a k-colouring of a graph, then I can easily check whether you are right. Thus the problem of k-colouring a graph is in $\mathcal{NP}$. On the other hand, finding a k-colouring can involve a lot of work, including much backtracking. It can be proved that this problem is $\mathcal{NP}$-complete and so equivalent to SAT. You might like to think about how graph colouring is related to solving Sudoku puzzles.

In proving that a problem is $\mathcal{NP}$-complete, one can make use of previously established results. So, if your problem is in the class $\mathcal{NP}$ and SAT can be suitably encoded into it, then your problem too must be $\mathcal{NP}$-complete, intuitively because any fast solution to your problem would lead to a fast solution to SAT. In the case of minesweeper, Richard Kaye actually shows how Boolean circuits (see Chapter 2) can be built as minesweeper elements.

This is not quite the end of the story. We don't know whether $\mathcal{P}$ is equal to $\mathcal{NP}$ but there are special cases where we know there are fast algorithms. For example, we saw in Section 1.8.3 that there is a fast algorithm for deciding whether a Horn formula is satisfiable or not. The point is that in the context of a particular problem the wff that arise may have extra properties and these might be susceptible to a fast method for determining satisfiability. In this case, whether $\mathcal{P}$ is equal to $\mathcal{NP}$ may actually be irrelevant.

---

[20] Only in the deluxe version of this book will you see actual colours.

Exercises 1.9

1. The picture below shows a graph.

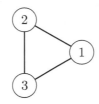

The vertices are to be coloured either blue or red under certain constraints that will be described by means of a wff in PL denoted by A. The following is a list of atoms and what they mean.

- $p =$ 'vertex 1 is blue'.
- $q =$ 'vertex 1 is red'.
- $r =$ 'vertex 2 is blue'.
- $s =$ 'vertex 2 is red'.
- $u =$ 'vertex 3 is blue'.
- $v =$ 'vertex 3 is red'.

Here is the wff A constructed from these atoms.

$$(p \oplus q) \wedge (r \oplus s) \wedge (u \oplus v) \wedge$$
$$(q \rightarrow (r \wedge u)) \wedge (p \rightarrow (s \wedge v)) \wedge$$
$$(s \rightarrow (p \wedge u)) \wedge (r \rightarrow (q \wedge v)) \wedge$$
$$(v \rightarrow (p \wedge r)) \wedge (u \rightarrow (q \wedge s))$$

(a) Translate A into pithy English.

(b) Use [76] to construct a truth table of A.

(c) Interpret this truth table.

1.10   Valid arguments

An argument is a connected series of statements intended to establish a proposition. — Monty Python.

We have so far viewed PL as a low-level description language. This is its significance in the question as to whether $\mathcal{P}$ equals $\mathcal{NP}$ or not. However, back in the Introduction, logic was described as the subject that dealt with reasoning. In fact, it is the subject that deals with the question of what it means to reason correctly. In this section, we shall show how the logic we have introduced so far can enable us to determine whether an argument is valid or not.

1.10.1 Definitions and examples

We begin with some simple examples of everyday reasoning.

Example 1.10.1. Someone makes the following announcement: 'In the pub either I will drink cider or I will drink apple juice. But I won't drink cider'. What do you deduce? Of course, you deduce that they will drink apple juice.

Example 1.10.2. If it is raining then I get wet. I am not wet. I deduce that it is not raining.

Example 1.10.3. The newspaper headline proclaims 'Either Smith or Jones will win the election. If Smith wins the election we are doomed. If Jones wins the election we are doomed'. What do you deduce? Clearly, with a sinking heart, you deduce that we are doomed.

None of these examples is at all exceptional. I would guess that in every case you had no trouble seeing that what we deduce *follows from* the preceding statements. But what do we mean by 'follows from'? This seems like something that would be impossible to define, like good art. In fact, we can use the logic we have described so far to explain why, in each case, it makes sense to say 'follows from'.

Let $A_1, \ldots, A_n, B$ be wff. An *argument* is by definition of the following form

$$A_1, \ldots, A_n \therefore B$$

where the wff $A_1, \ldots, A_n$ are called the *assumptions* or *premisses*,[21] the wff B is called the *conclusion* and the symbol '$\therefore$' is read 'therefore' which is another way of saying 'follows from'. Thus Example 1.10.1 could be written in the following way:

Either I will drink cider or I will drink apple juice. I won't drink cider $\therefore$ I will drink apple juice.

Anyone can claim they are making an argument, so the question is how can

---

[21]This is the usual spelling used for the notion in logic. The alternative, and equally correct, spelling is 'premise'. At this point, I am obliged to tell the story about the feuding neighbours who could never agree because they were arguing from different premises. This is a variant of a *bon mot* attributed to the Reverend Sydney Smith, amongst others.

we use logic to decide whether their claim is correct or not? To do this, we introduce the relation $\models$ which is defined as follows:

$$A_1, \ldots, A_n \models B$$

if whenever all of $A_1, \ldots, A_n$ are true then B must be true as well. Equivalently, it is impossible for $A_1, \ldots, A_n$ to be true and B to be false. We now define the argument

$$A_1, \ldots, A_n \therefore B$$

to be *valid* precisely when

$$A_1, \ldots, A_n \models B.$$

Books on logic used in philosophy go to some lengths to explain the difference in meaning between $\therefore$ and $\models$ [66, Section 13.7]. Mathematicians tend to be cavalier about these things and consequently in this book $\models$ can also be read as meaning 'therefore', as well.

The definition of a valid argument encapsulates many important examples of logical reasoning, but we shall see later that there are also examples of logical reasoning that cannot be captured by PL. It is this that will lead us to the generalization of PL called *first-order logic*, or FOL.

Examples 1.10.4. Here are some examples of valid arguments.

1. $p \vee q, \neg p \models q$. We show this is a valid argument.

p	q	$\neg p$	$p \vee q$
T	T	F	T
T	F	F	T
F	T	T	T
F	F	T	F

We are only interested in the cases where both $p \vee q$ and $\neg p$ are true.

p	q	$\neg p$	$p \vee q$
F	T	T	T

We see that if $p \vee q$ and $\neg p$ are true then q must be true. This, of course, is the general form of the argument used in Example 1.10.1 where p is the statement 'I will drink cider' and q is the statement 'I will drink apple juice'.

2. $p \rightarrow q, \neg q \models \neg p$. We show this is a valid argument.

p	q	$\neg p$	$\neg q$	$p \rightarrow q$
T	T	F	F	T
T	F	F	T	F
F	T	T	F	T
F	F	T	T	T

We are only interested in the cases where both $p \to q$ and $\neg q$ are true.

p	q	$\neg p$	$\neg q$	$p \to q$
F	F	T	T	T

We see that if $p \to q$ and $\neg q$ are true then $\neg p$ must be true. This, of course, is the general form of the argument used in Example 1.10.2 where p is the statement 'it is raining' and q is the statement 'I get wet'.

3. $p \vee q, p \to r, q \to r \vDash r$. We show that this is a valid argument. The table below shows all the truth value combinations we need.

p	q	r	$p \vee q$	$p \to r$	$q \to r$
T	T	T	T	T	T
T	T	F	T	F	F
T	F	T	T	T	T
T	F	F	T	F	T
F	T	T	T	T	T
F	T	F	T	T	F
F	F	T	F	T	T
F	F	F	F	T	T

But we are only interested in the cases where $p \vee q$ and $p \to r$ and $q \to r$ are all true. Therefore, we only need the following:

p	q	r	$p \vee q$	$p \to r$	$q \to r$
T	T	T	T	T	T
T	F	T	T	T	T
F	T	T	T	T	T

However, you can now see that r is true in every case.

This, of course, is the general form of the argument used in Example 1.10.3 where p is the statement 'Smith wins the election', q is the statement 'Jones wins the election' and r is the statement 'we are doomed'.

4. $p, p \to q \vDash q$. We show that this is a valid argument. Here is the truth table for $p \to q$.

p	q	$p \to q$
T	T	T
T	F	F
F	T	T
F	F	T

We are only interested in the cases where both p and $p \to q$ are true.

p	q	$p \to q$
T	T	T

We see that if p and $p \to q$ are true then q must be true. Thus the argument is valid.

5. $p \to q, q \to r \vDash p \to r$. We show this is a valid argument.

p	q	r	$p \to q$	$q \to r$	$p \to r$
T	T	T	T	T	T
T	T	F	T	F	F
T	F	T	F	T	T
T	F	F	F	T	F
F	T	T	T	T	T
F	T	F	T	F	T
F	F	T	T	T	T
F	F	F	T	T	T

We are only interested in the cases where both $p \to q$ and $q \to r$ are true.

p	q	r	$p \to q$	$q \to r$	$p \to r$
T	T	T	T	T	T
F	T	T	T	T	T
F	F	T	T	T	T
F	F	F	T	T	T

We see that in every case $p \to r$ is true. Thus the argument is valid.

We may always use truth tables in the way we did above to decide whether an argument is valid or not, but it is quite laborious. The following result, however, enables us to do this in a more straightforward way.

Proposition 1.10.5. *The argument*

$$A_1, \ldots, A_n \vDash B$$

is valid precisely when

$$\vDash (A_1 \wedge \ldots \wedge A_n) \to B.$$

Proof. The result is proved in two steps.

1. We prove that $A_1, \ldots, A_n \vDash B$ precisely when $A_1 \wedge \ldots \wedge A_n \vDash B$. This means that we can restrict to the case of exactly one wff on the left-hand side of the semantic turnstile. Suppose first that $A_1, \ldots, A_n \vDash B$ is a valid argument and that it is not the case that $A_1 \wedge \ldots \wedge A_n \vDash B$ is a valid argument. Then there is some assignment of truth values to the atoms that makes $A_1 \wedge \ldots \wedge A_n$ true and B false. But this means that each of $A_1, \ldots, A_n$ is true and B is false, which contradicts the fact that we are given $A_1, \ldots, A_n \vDash B$ is a valid argument. Suppose now that $A_1 \wedge \ldots \wedge A_n \vDash B$ is a valid argument but that $A_1, \ldots, A_n \vDash B$ is not

a valid argument. Then some assignment of truth values to the atoms makes $A_1, \ldots, A_n$ true and B false. This means that $A_1 \wedge \ldots \wedge A_n$ is true and B is false. But this contradicts that we are given that $A_1 \wedge \ldots \wedge A_n \vDash B$ is a valid argument.

2. We prove that $A \vDash B$ precisely when $\vDash A \to B$. Suppose first that $A \vDash B$ is a valid argument and that $A \to B$ is not a tautology. Then there is some assignment of the truth values to the atoms that makes A true and B false. But this contradicts that $A \vDash B$ is a valid argument. Suppose now that $A \to B$ is a tautology and $A \vDash B$ is not a valid argument. Then there is some assignment of truth values to the atoms that makes A true and B false. But then the truth value of $A \to B$ is false, from the way implication is defined, and this is a contradiction. $\qquad\square$

Once again, we see the important role tautologies play in PL.

Example 1.10.6. We show that $p, p \to q \vDash q$ is a valid argument by showing that $\vDash (p \wedge (p \to q)) \to q$ is a tautology.

p	q	$p \to q$	$p \wedge (p \to q)$	$(p \wedge (p \to q)) \to q$
T	T	T	T	T
T	F	F	F	T
F	T	T	F	T
F	F	T	F	T

1.10.2   Proof in mathematics (I)

Each of the arguments described in Examples 1.10.4 seems quite simple, possibly even simple-minded. But if such arguments are combined, results of great complexity can be obtained. This is the basic idea behind the notion of mathematical proof. Proofs are, in fact, the essence of mathematics and distinguish it from any other subject. The idea of proofs arose first in classical Greece, whereas the logic described in this book grew out of attempts in the early decades of the twentieth century to provide a proper definition of what a proof was.[22] To conclude this section, I shall describe a few examples of mathematical proofs where the ideas of this section can be seen in action. For more background information on proofs, see the first two chapters of [44], and for a more thorough treatment of how logic is used in mathematics, see [15]. I shall return to this topic in Section 3.3. The specific proof technique known as proof by induction will be described in Section 2.1.

---

[22]The notion of proof is so powerful that it has been borrowed by computer scientists since it is needed to answer the most basic question we can ask about a program which is whether it does what we claim it does. It is certainly not enough just to test the program with a few trial inputs; think, for example, of the programs that help modern aircraft fly-by-wire. You would like to know that those programs will always work, not merely in the cases that someone thought to check.

In the Introductory exercises, I included a question about Sudoku. This may have seemed odd, but in Sections 1.5 and 1.9 I explained that these puzzles are prototypes for a wide class of problems known as $\mathcal{NP}$-complete. I now want to look at Sudoku from another perspective: that of a toy model for how mathematics works. This will lead us to a more 'dynamic' way of thinking about logic which has, at its core, the idea of an *argument*. Sudoku puzzles are solved as follows. First, you study the initial information you have been given; that is, the squares that have already been filled in with numbers. To solve the puzzle, you have to accept these as a given. Next, you use reason to try to fill in other squares in accordance with the rules of the game. Sometimes, filling a square with a number leads to a conflict, or as we would say in mathematics a *contradiction*, so that you know your choice of number was incorrect. This process is 'dynamic' rather than the 'static' approach adopted in Section 1.5 via the satisfiability problem.

My description for how a Sudoku puzzle is solved is also a good metaphor for how mathematics itself works. Each domain of mathematics is character- ized by its basic assumptions or *axioms*. These are akin to the initial informa- tion given in a Sudoku puzzle and the rules of Sudoku. Next, by using reason and appealing to previously proved results, you attempt to prove new results.

Logic is the very grammar of reason. But just as a knowledge of Russian grammar won't make you a Tolstoy, so too a knowledge of logic won't make you a Gauss. If you want, however, to do mathematics, then you have to acquire an understanding of logic, just as if you want to write you have to acquire an understanding of grammar. We conclude with a few examples of how the logic we have developed so far can be used in mathematics.

How would you prove a statement of the form $p \to q$? We know that the only case where this statement is false is where p is true and q is false. So, it is enough to assume that p is true and prove that q has to be true. However, this is not quite the whole story. In mathematics, we are always working with arguments of the form $\Gamma, p \vdash q$ where Γ is the set of all mathematical truths relevant to the mathematical domain in question and any axioms that need to be assumed to work in that domain. This should always be borne in mind in what follows. An illustration of this approach is described in the following example.

Example 1.10.7. Let n be a natural number. So, $n = 0, 1, 2, \ldots$. We say it is *even* if it is exactly divisible by two. We say that n is *odd* if '$\neg$ (n is even)'. This definition is not very useful and a better, but equivalent, one is to say that n is odd if the remainder 1 is left when n is divided by 2. I prove first that the square of an even number is even and the square of an odd number is odd. Let n be even. Then by definition, we can write $n = 2m$. Squaring both sides we obtain $n^2 = 4m^2 = 2(2m^2)$. It follows that n^2 is even. Now, let n be odd. By definition, we can write $n = 2m + 1$. Squaring both sides we obtain $n^2 = 4m^2 + 4m + 1 = 2(2m^2 + 2m) + 1$. It follows that n^2 is odd.

Our next example illustrates an argument of the following form

$$p \rightarrow q, \neg q \vDash \neg p.$$

Example 1.10.8. We prove the *converse* of Example 1.10.7. Suppose I tell you that n^2 is even. What can you deduce about n? Hopefully, you can see, intuitively, that n is even, but we can use what we have learnt in this section to prove it. Let p be the statement 'n is odd'. Let q be the statement 'n^2 is odd'. Thus $\neg q$ is the statement that 'n^2 is even' and $\neg p$ is the statement that 'n is even'. The following argument is valid

$$p \rightarrow q, \neg q \vDash \neg p.$$

Thus the following argument is valid

n odd implies n^2 is odd, n^2 is even therefore n is even.

But we have proved that 'n odd implies n^2 is odd' is always true and I have told you that n^2 is even. It follows that n is even. We have therefore proved that if the square of a natural number is even then that natural number is also even.

Suppose we are trying to prove that the statement p follows from some set of assumptions Γ. One way to do this is to construct a valid argument of the form $\Gamma, \neg p \vDash \boldsymbol{f}$, where you can think of $\boldsymbol{f}$ as being any contradiction. This is equivalent to proving $\Gamma \vDash \neg p \rightarrow \boldsymbol{f}$. But the only way for the statement $\neg p \rightarrow \boldsymbol{f}$ to be true is if $\neg p$ is false which implies that p is true. This is called *proof by contradiction*, a method of proof that often leads to profound results, as illustrated by the following example.

Example 1.10.9. Define a real number to be *rational* if it can be written as a fraction $\frac{a}{b}$. A real number that is *not* rational is called *irrational*. Observe that the word 'irrational' is really 'in-rational' where 'in' just means 'not'.[23] Rational numbers are 'graspable' or 'intelligible' since they can be described by two natural numbers; this is why they are 'rational' (the more usual meaning of that word). Irrational numbers are just the opposite. I shall prove that $\sqrt{2}$ is irrational; I shall explain why this is interesting afterwards.

Let p be the statement '$\sqrt{2} = \frac{a}{b}$ where a and b are natural numbers'. Let q be the statement 'a and b are both even'. I shall first prove the statement $p \rightarrow q$. To do this, we assume that p is true and deduce that q is true.

1. Assume p is true. That is, assume that $\sqrt{2} = \frac{a}{b}$ for some natural numbers a and b.

2. Square both sides of the equation $\sqrt{2} = \frac{a}{b}$ to obtain $2 = \frac{a^2}{b^2}$. Then rearrange to obtain $2b^2 = a^2$. I can do this by using previously proved results in mathematics (and so this is part of our background assumptions Γ).

---

[23] In the same way that 'impossible' is 'in-possible'. The way the letter 'n' changes in each case is an example of *assimilation*.

3. a^2 is even by definition.

4. a is even by the result we proved in Example 1.10.9.

5. $a = 2x$ for some natural number x by definition of the word 'even'.

6. It follows that $b^2 = 2x^2$ using previously proved results in mathematics.

7. b^2 is even by definition.

8. b is even by the result we proved in Example 1.10.9.

9. Since a and b are both even it follows that the statement q is true.

The above argument did not depend on a and b apart from the fact that $\sqrt{2} = \frac{a}{b}$.

We now prove that we can always assume that at least one of a or b is odd. The idea is that if both a and b are even then we can divide out by the largest power of 2 that divides them both. Thus if $a = 2^m a'$ and $b = 2^n b'$, where 2^m is the largest power of 2 that divides a and where 2^n is the largest power of 2 that divides b, then the following cases arise:

$$\frac{a}{b} = \frac{a'}{b'}$$

if $m = n$, where both a' and b' are odd;

$$\frac{a}{b} = \frac{a'}{2^{n-m} b'}$$

if $n > m$, where a' is odd and $2^{n-m} b'$ is even; and

$$\frac{a}{b} = \frac{2^{m-n} a'}{b'}$$

if $n < m$, where $2^{m-n} a'$ is even and b' is odd.

Observe that $\neg q$ is logically equivalent to the statement 'at least one of a and b is odd' because

$$\neg((a \text{ is even }) \wedge (b \text{ is even })) \equiv (a \text{ is odd}) \vee (b \text{ is odd}).$$

Thus we can assume that $\neg q$ is true by choosing a and b appropriately but, as before, we can then prove that $p \to q$ is true. Under these assumptions $p \to (q \wedge \neg q)$ is also true because

$$p \to s, p \to t \vDash p \to (s \wedge t)$$

can be checked to be a valid argument. But $q \wedge \neg q$ is always false. Thus p must be false and so $\neg p$ is true. It follows that $\sqrt{2}$ is irrational.

Why should you care about this result? The reason is that it implies that you can never calculate an exact rational value for $\sqrt{2}$. If you calculate $\sqrt{2}$

on your calculator the number that appears is in fact a rational number. For example, my calculator tells me that

$$\sqrt{2} = 1 \cdot 414213562$$

but this is just the rational number

$$\frac{1414213562}{1000000000}$$

in disguise. We proved above that this cannot be the exact value of $\sqrt{2}$. Despite this, my calculator believes that when you square $1 \cdot 414213562$ you get 2. The moral is that you cannot always believe what your calculator tells you.

Exercises 1.10

---

1. Determine which of the following really are valid arguments.

   (a) $p \rightarrow q \vDash \neg q \vee \neg p$.

   (b) $p \rightarrow q, \neg q \rightarrow p \vDash q$.

   (c) $p \rightarrow q, r \rightarrow s, p \vee r \vDash q \vee s$.

   (d) $p \rightarrow q, r \rightarrow s, \neg q \vee \neg s \vDash \neg p \vee \neg r$.

2. This question introduces the idea of *resolution* important in some computer science applications of logic. Let x_1, x_2, x_3 and y_2, y_3 be two sets of literals. Show that the following is a valid argument

   $$(x_1 \vee x_2 \vee x_3), (\neg x_1 \vee y_2 \vee y_3) \vDash x_2 \vee x_3 \vee y_2 \vee y_3.$$

3. This question explores some of the properties of the symbol $\vDash$ and will be used in Section 1.12. First, we extend the notation a little. We define

   $$A_1, \ldots, A_m \vDash B_1, \ldots, B_n$$

   to mean the same thing as

   $$A_1, \ldots, A_m \vDash B_1 \vee \ldots \vee B_n.$$

   (a) Show that $A_1, \ldots, A_m, X \vDash B_1, \ldots, B_n, X$ is always a valid argument.

   (b) Show that if $A_1, \ldots, A_m, X, Y \vDash B_1, \ldots, B_n$ is a valid argument so too is $A_1, \ldots, A_m, X \wedge Y \vDash B_1, \ldots, B_n$.

   (c) Show that if $A_1, \ldots, A_m \vDash B_1, \ldots, B_n, X, Y$ is a valid argument so too is $A_1, \ldots, A_m \vDash B_1, \ldots, B_n, X \vee Y$.

(d) Show that if $A_1, \ldots, A_m, X \vDash B_1, \ldots, B_n$ is a valid argument so too is $A_1, \ldots, A_m \vDash \neg X, B_1, \ldots, B_n$.

(e) Show that if $A_1, \ldots, A_m \vDash B_1, \ldots, B_n, X$ is a valid argument so too is $A_1, \ldots, A_m, \neg X \vDash B_1, \ldots, B_n$.

(f) Show that if

$$A_1, \ldots, A_m \vDash B_1, \ldots, B_n, X \text{ and } A_1, \ldots, A_m \vDash B_1, \ldots, B_n, Y$$

are valid arguments so too is

$$A_1, \ldots, A_m \vDash B_1, \ldots, B_n, X \wedge Y.$$

(g) Show that if

$$A_1, \ldots, A_m, X \vDash B_1, \ldots, B_n \text{ and } A_1, \ldots, A_m, Y \vDash B_1, \ldots, B_n$$

are valid arguments so too is

$$A_1, \ldots, A_m, X \vee Y \vDash B_1, \ldots, B_n.$$

(h) Show that if

$$A_1, \ldots, A_m \vDash B_1, \ldots, B_n, X \text{ and } A_1, \ldots, A_m, Y \vDash B_1, \ldots, B_n$$

are valid arguments so too is

$$A_1, \ldots, A_m, X \to Y \vDash B_1, \ldots, B_n.$$

1.11 Truth trees

All problems about PL can be answered using truth tables. But there are two problems:

1. The method of truth tables is hard work.

2. The method of truth tables does not generalize to first-order logic.

In this section, we shall describe an algorithm that is often more efficient than truth tables but which, more significantly, can also be generalized to first-order logic. This is the method of truth trees. It will often speed up the process of determining the following:

- Deciding whether a wff (or set of wff) is satisfiable or not.

- Deciding whether a wff is a tautology or not.

- Deciding whether an argument is valid or not.

- Converting a wff into DNF.

It is based on the following ideas:

- Given a wff A that we wish to determine is satisfiable or not, we start by assuming A is satisfiable and work backwards.

- We use a data structure, a tree, to keep track efficiently of all possibilities that occur.

- We break A into smaller pieces (and so the algorithm is of a type known as *divide and conquer*).

- What is not true is false; thus we need only keep track of what is true and any other cases will automatically be false.

Truth trees are often called *(semantic) tableaux* in the literature.

1.11.1 The truth tree algorithm

The starting point is to consider the various possible shapes that a wff A can have. Observe that since $X \oplus Y \equiv \neg(X \leftrightarrow Y)$, I shall therefore not mention $\oplus$ explicitly in what follows. There are therefore nine possibilities for A:

$X \wedge Y$	$X \vee Y$	$X \to Y$
$X \leftrightarrow Y$	$\neg(X \wedge Y)$	$\neg(X \vee Y)$
$\neg(X \to Y)$	$\neg(X \leftrightarrow Y)$	$\neg\neg X$

We now introduce a graphical way of representing the truth values of A in terms of the truth values for X and Y. We introduce two kinds of graphical rules. I shall give the *forms* of the two rules first and then the exact rules we shall be using.

- The α-*rule* is non-branching/and-like. If α is a wff then this rule looks like

This means that a truth assignment satisfies α precisely when it satisfies both α_1 and α_2.

- The *β-rule* is branching/or-like. If β is a wff then this rule looks like

$$\beta$$
$$\overbrace{\beta_1 \quad \beta_2}$$

This means that a truth assignment satisfies β precisely when it satisfies at least one of β_1 or β_2.

In the above two rules, we call α_1 and α_2 the *successors* of α, and likewise for β_1 and β_2 with respect to β.

We now list the truth tree rules for the nine forms of the wff given above.

α-rules

$X \wedge Y$	$\neg(X \vee Y)$	$\neg(X \to Y)$	$\neg\neg X$
|	|	|	|
X	$\neg X$	X	X
Y	$\neg Y$	$\neg Y$	

A wff A which is equal to one of the wff $X \wedge Y$, $\neg(X \vee Y)$, $\neg(X \to Y)$ and $\neg\neg X$ is said to be an *α-formula*. Sometimes, we just refer to a formula α, meaning an α-formula. In this case, we refer to the formulae α_1 and α_2 being the two formulae that arise in applying an α-rule — though in the case where the α-formula is $\neg\neg X$ both α_1 and α_2 are equal to X.

β-rules

$X \vee Y$	$\neg(X \wedge Y)$	$X \to Y$	$X \leftrightarrow Y$	$\neg(X \leftrightarrow Y)$
$X \quad Y$	$\neg X \quad \neg Y$	$\neg X \quad Y$	$X \quad \neg X$	$X \quad \neg X$
			$Y \quad \neg Y$	$\neg Y \quad Y$

A wff A which is equal to one of the wff $X \vee Y$, $\neg(X \wedge Y)$, $X \to Y$, $X \leftrightarrow Y$ and $\neg(X \leftrightarrow Y)$ is said to be a *β-formula*. Sometimes, we just refer to a formula β, meaning a β-formula. In this case, we refer to the formulae β_1 and β_2 being the two formulae that arise in applying a β-rule — though in the case where the β-formula are $X \leftrightarrow Y$ and $\neg(X \leftrightarrow Y)$ each of β_1 and β_2 consists of two formulae.

> The α- and β-rules listed above are easy to remember if you bear in mind that they simply encode in tree form the truth tables for the respective connectives.

A truth tree for a wff A will be constructed by combining the small trees above in such a way that we shall be able to determine when A is satisfiable and how. The following definition dealing with trees is essential. Recall that a *branch* of a tree is a path that starts at the root and ends at a leaf. For example in the tree below

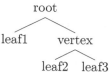

root

leaf1   vertex

leaf2   leaf3

there are three branches. The key idea in what follows can now be stated:

truth flows down the branches, and different branches represent alternative possibilities and so should be regarded as being combined by means of disjunctions.

I shall describe the full truth tree algorithm once I have worked my way through some illustrative examples.

Don't confuse parse trees *which are about syntax with* truth trees *which are about semantics.*

Example 1.11.1. Find all truth assignments that make the wff $A = \neg p \to (q \wedge r)$ true using truth trees. The first step is to place A at the root of what will be the truth tree

$$\neg p \to (q \wedge r).$$

We now use the branching rule for $\to$ to get

$$\neg p \to (q \wedge r)\checkmark$$

$$\neg(\neg p) \quad q \wedge r$$

I have used the check symbol $\checkmark$ to indicate that I have *used* the occurrence of the wff $\neg p \to (q \wedge r)$. It is best to be systematic and so I shall work from left-to-right. We now apply the truth tree rule for double negation to get

$$\neg p \to (q \wedge r)\checkmark$$

$$\checkmark \neg(\neg p) \quad q \wedge r$$

$$p$$

where once again I have used the check symbol $\checkmark$ to indicate that that occurrence of a wff has been used.

Finally, I apply the truth tree rule for conjunction to get

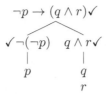

$$\neg p \to (q \land r)\checkmark$$

There are now no further truth tree rules I can apply and so we say that the truth tree is *finished*. We deduce that the root A is true precisely when either p is true or q and r are both true. In other words, $A \equiv p \lor (q \land r)$, which is in DNF. You can easily check that this contains exactly the same information as the rows of the truth table of A that output T. By our idea above we know that all other truth values must be F.

Before giving some more complex examples, let me highlight some important points (all of which can be proved).

- It is the branches of the truth tree that contain information about the truth table. Each branch contains information about one or more rows of the truth table.

- It follows that all the literals on a branch must be true.

- Thus if an atom and its negation occur on the same branch then there is a contradiction. That branch is then *closed* by placing the cross symbol, ✗, at its leaf. No further growth takes place at a closed leaf.

The next example illustrates an important point about applying the rules.

Example 1.11.2. Find all the truth assignments that satisfy

$$\neg((p \lor q) \to (p \land q)).$$

Here is the truth tree for this wff.

$$\checkmark \neg((p \lor q) \to (p \land q))$$
$$\checkmark p \lor q$$
$$\checkmark \neg(p \land q)$$

$$p \qquad\qquad q$$

$$\text{✗} \neg p \quad \neg q \qquad \neg p \quad \neg q \text{ ✗}$$

The truth tree is finished and there are two open branches (those branches not marked with a ✗). The first branch tells us that p and $\neg q$ are both true and the second tells us that $\neg p$ and q are both true. It follows that the wff has the DNF

$$(p \land \neg q) \lor (\neg p \land q).$$

76

The key point to observe in this example is that when I applied the β-rule to the wff $\neg(p \wedge q)$ *I applied it to all branches that contained that wff*. This is crucially important since *possibilities multiply*.

The following diagrams illustrate the general idea contained in the above example. A truth tree with the following form

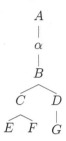

becomes

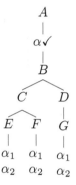

after the application of the α-rule to the wff α, and a truth tree with the following form

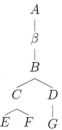

becomes

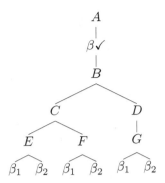

after the application of the β-rule to the wff β.

The above examples also suggest the following *strategy*:

> Apply α-rules before β-rules since the application of β-rules leads to the tree gaining more branches, and subsequent applications of any rule must be appended to every branch.

Here are some further examples which should be worked through carefully.

Example 1.11.3. Find all satisfying truth assignments to

$$(p \wedge (q \to r)) \vee (\neg p \wedge (r \to q)).$$

Here is the truth tree of this wff:

$$(p \wedge (q \to r)) \vee (\neg p \wedge (r \to q)) \checkmark$$

$$\checkmark p \wedge (q \to r) \qquad \neg p \wedge (r \to q) \checkmark$$

$$p \qquad\qquad \neg p$$

$$\checkmark q \to r \qquad\qquad r \to q \checkmark$$

$$\neg q \quad r \qquad\qquad \neg r \quad q$$

There are four branches and all branches are open. These lead to the DNF

$$(p \wedge \neg q) \vee (p \wedge r) \vee (\neg r \wedge p) \vee (q \wedge \neg p).$$

The satisfying truth assignments can now easily be found.

Example 1.11.4. Write

$$(p \rightarrow (q \rightarrow r)) \rightarrow ((p \rightarrow q) \rightarrow r)$$

in DNF. Here is the truth tree of this wff:

$$(p \rightarrow (q \rightarrow r)) \rightarrow ((p \rightarrow q) \rightarrow r)\checkmark$$

```
                      (p → (q → r)) → ((p → q) → r)✓
                     ╱                              ╲
       ✓¬(p → (q → r))                    (p → q) → r ✓
              |                              ╱        ╲
              p                        ¬(p → q)✓      r
       ✓¬(q → r)                           |
              |                            p
              q                           ¬q
             ¬r
```

There are three branches all open. These lead to the DNF

$$(p \wedge q \wedge \neg r) \vee (p \wedge \neg q) \vee r.$$

Example 1.11.5. Write

$$[(p \wedge \neg q) \rightarrow (q \wedge r)] \rightarrow (s \vee \neg q),$$

which contains four atoms, in DNF. Here is the truth tree of this wff:

$$((p \wedge \neg q) \rightarrow (q \wedge r)) \rightarrow (s \vee \neg q)\checkmark$$

```
                 ((p ∧ ¬q) → (q ∧ r)) → (s ∨ ¬q)✓
                ╱                               ╲
   ✓¬((p ∧ ¬q) → (q ∧ r))              s ∨ ¬q ✓
              |                          ╱    ╲
        ✓ p ∧ ¬q                        s     ¬q
        ✓¬(q ∧ r)
              |
              p
             ¬q
            ╱  ╲
          ¬q    ¬r
```

There are four branches all open. These lead to the DNF

$$(p \wedge \neg q) \vee (\neg r \wedge \neg q \wedge p) \vee s \vee \neg q.$$

It's important to remember that the method of truth trees is an algorithm for finding those truth assignments that make a wff true. This leads to the following important result, which should be studied carefully, since it is the source of error and confusion.

Proposition 1.11.6. *To show that X is a tautology show that the truth tree for $\neg X$ has the property that every branch closes.*

Proof. If the truth tree for $\neg X$ has the property that every branch closes then $\neg X$ is not satisfiable. This means that $\neg X$ is a contradiction. Thus X is a tautology. $\square$

Here are some examples of showing that a wff is a tautology.

Example 1.11.7. Determine whether

$$X = ((p \to q) \land (p \to r)) \to (p \to (q \land r))$$

is a tautology or not. We begin the truth tree with $\neg X$.

The tree for $\neg X$ closes and so $\neg X$ is a contradiction. Thus X is a tautology.

Example 1.11.8. Determine whether

$$X = (p \to (q \to r)) \to ((p \to q) \to (p \to r))$$

is a tautology or not.

The tree for $\neg X$ closes and so $\neg X$ is a contradiction. Thus X is a tautology.

We can also use truth trees to determine whether an argument is valid or not. To see how, we use Proposition 1.10.5, which showed that $A_1, \ldots, A_n \vDash B$ precisely when $\vDash (A_1 \wedge \ldots \wedge A_n) \to B$. Thus to show that an argument is valid using a truth tree, we could place $\neg[(A_1 \wedge \ldots \wedge A_n) \to B]$ at its root. But this is logically equivalent to $A_1 \wedge \ldots \wedge A_n \wedge \neg B$. Thus we need only list $A_1, \ldots, A_n, \neg B$ at the root of our truth tree. If every branch closes the corresponding argument is valid, and if there are open branches then the argument is not logically valid.

Examples 1.11.9. We use truth trees to show that our simple examples of arguments really are valid.

1. $p, p \to q \vDash q$ is a valid argument.[24]

$$
\begin{array}{c}
p \\
p \to q \ \checkmark \\
\neg q \\
\diagdown \\
\neg p \ \textbf{X} \quad q \ \textbf{X}
\end{array}
$$

The tree closes and so the argument is valid.

2. $p \to q, \neg q \vDash \neg p$ is a valid argument.[25]

$$
\begin{array}{c}
p \to q \ \checkmark \\
\neg q \\
\neg\neg p \ \checkmark \\
| \\
p \\
\diagdown \\
\neg p \ \textbf{X} \quad q \ \textbf{X}
\end{array}
$$

The tree closes and so the argument is valid.

3. $p \vee q, \neg p \vDash q$ is a valid argument.[26]

$$
\begin{array}{c}
p \vee q \ \checkmark \\
\neg p \\
\neg q \\
\diagdown \\
p \ \textbf{X} \quad q \ \textbf{X}
\end{array}
$$

The tree closes and so the argument is valid.

4. $p \to q, q \to r \vDash p \to r$ is a valid argument.[27]

---

[24] Known as *modus ponens*. The phrase in italics is the Latin name for this particular valid argument; I have given the corresponding Latin names for the other valid arguments in the next three footnotes. You do not have to learn these phrases. Philosophers might know them but mathematicians assuredly don't, except possibly modus ponens, though they have no idea what that Latin phrase actually means any more than I do unless I look it up.

[25] Known as *modus tollens*.

[26] Known as *disjunctive syllogism*.

[27] Known as *hypothetical syllogism*.

$$p \to q \checkmark$$
$$q \to r \checkmark$$
$$\neg(p \to r) \checkmark$$
$$|$$
$$p$$
$$\neg r$$

$\neg p$ ✗ $\qquad$ q

$\neg q$ ✗ $\quad$ r ✗

The tree closes and so the argument is valid.

Example 1.11.10. Show that the following

$$(\neg q \to \neg p) \wedge (\neg r \to \neg q), s \wedge (s \to \neg r), t \to p \vDash \neg t$$

is a valid argument.

$$(\neg q \to \neg p) \wedge (\neg r \to \neg q) \checkmark$$
$$s \wedge (s \to \neg r) \checkmark$$
$$t \to p \checkmark$$
$$\neg\neg t \checkmark$$
$$|$$
$$t$$
$$|$$
$$\neg q \to \neg p \checkmark$$
$$\neg r \to \neg q \checkmark$$
$$|$$
$$s$$
$$s \to \neg r \checkmark$$

$\neg t$ ✗ $\qquad\qquad$ p

$\neg\neg q$ $\qquad$ $\neg p$ ✗

q

$\neg\neg r$ $\qquad$ $\neg q$ ✗

r

$\neg s$ ✗ $\quad$ $\neg r$ ✗

The tree closes and so the argument is valid.

Truth tree algorithm

Input: wff X.

Procedure: Place X at what will be the root of the truth tree. Depending on its shape, apply either an α-rule or a β-rule. Place a ✓against X to indicate that it has been *used*.

Now repeat the following:

- *Close* any branch that contains an atom and its negation by placing a cross ✗ beneath the leaf defining that branch.

- If all branches are closed then *stop* since the tree is now *finished and closed*.

- If not all branches are closed but the tree contains only literals or used wff then *stop* since the tree is now *finished and open*.

- If the tree is not finished then choose an unused wff Y which is not a literal. Now do the following: for each open branch that contains Y append the effect of applying either the α-rule or β-rule to Y, depending on which is appropriate, to the leaf of that branch.

Output:

- If the tree is finished and closed then X is a contradiction.

- If the tree is finished and open then X is satisfiable. We may find all the truth assignments that make X true as follows: for each open branch in the finished tree for X assign the value T to all the literals in that branch. If any atoms are missing from this branch then they may be assigned truth values arbitrarily.

Applications of truth trees

1. To prove that X is *satisfiable*, show that the finished truth tree for X is open.

2. We say that a set of wff $A_1, \ldots, A_n$ is *satisfiable* if there is at least one single truth assignment that makes them all true. This is equivalent to saying that $A_1 \wedge \ldots \wedge A_n$ is satisfiable. Thus to show that $A_1, \ldots, A_n$ is satisfiable simply list these wff as the root of the truth tree.

3. To put X into *DNF*, construct the truth tree for X. Assume that when finished it is open. For each open branch i, construct the conjunction of the literals that appear on that branch C_i. Then form the disjunction of the C_i.

4. To prove that X is a *tautology*, show that the finished truth tree for $\neg X$ is closed.

5. To prove that $A_1, \ldots, A_n \models B$ is a *valid argument*, place

$$A_1, \ldots, A_n, \neg B$$

at the root of the tree and show that the finished truth tree is closed.

1.11.2 The theory of truth trees

In this section, we shall prove that truth trees really do the job we have claimed for them. But before we can prove our main theorem, we prove two lemmas, Lemma 1.11.13 and Lemma 1.11.15, that do all the heavy lifting for us. Some extra notation will be needed. Let X be a wff with atoms $p_1, \ldots, p_n$. Assigning truth values to these atoms will ultimately lead to a truth value being assigned to X. If we call the truth assignment to these atoms τ then the corresponding truth value of X is denoted by $\tau(X)$.

Example 1.11.11. Let $X = \neg((p \wedge q) \to r)$ and let τ be the following truth assignment

p	q	r
T	T	F

Then $\tau(p \wedge q) = T$ and $\tau((p \wedge q) \to r) = F$ so that $\tau(X) = T$.

The first result we shall prove states that if a finished truth tree with root X has an open branch then there is a truth assignment, constructed from that branch, that satisfies X. It is well to begin with a concrete example.

Example 1.11.12. We return to the truth tree for the wff $\neg((p \vee q) \to (p \wedge q))$.

$$\checkmark \neg((p \vee q) \to (p \wedge q))$$
$$|$$
$$\checkmark p \vee q$$
$$\checkmark \neg(p \wedge q)$$

$$p \qquad\qquad q$$

$$\text{✗} \neg p \quad \neg q \qquad \neg p \quad \neg q \text{ ✗}$$

The truth tree is finished and there are two open branches (those branches not marked with a ✗). Here are the sets of formulae in each of the two open branches

$$H_1 = \{\neg((p \vee q) \to (p \wedge q)), p \vee q, \neg(p \wedge q), p, \neg q\}$$

and

$$H_2 = \{\neg((p \vee q) \to (p \wedge q)), p \vee q, \neg(p \wedge q), q, \neg p\}.$$

We focus on the first set H_1, though what we shall say applies equally well to H_2. Observe that the set H_1 does not contain an atom and the negation of that atom. This is because H_1 arises from an open branch. Next observe, that if an α-wff is in H_1 then both α_1 and α_2 are also in H_1. In this instance, the only α-wff is $\neg((p \vee q) \rightarrow (p \wedge q))$ and here $\alpha_1 = p \vee q$ and $\alpha_2 = \neg(p \wedge q)$, both of which belong to H_1. Finally, if a β-wff belongs to H_1 then at least one of β_1 or β_2 belongs to H_1. In this case, there are two β-wff in H_1: namely, the wff $p \vee q$ and $\neg(p \wedge q)$. From the first wff the β_1-wff, p, belongs to H_1, and from the second wff the β_2-wff, $\neg q$, belongs to H_1. So, why are these properties of H_1 significant? Assign p the truth value T and assign q the truth value F. Call this the *initial truth assignment*. Now we 'climb the ladder': it follows that both p and $\neg q$ are true; it therefore follows that $p \vee q$ and $\neg(p \wedge q)$ are true; it therefore follows that $\neg((p \vee q) \rightarrow (p \wedge q))$ is true. Thus our initial truth assignment leads to every wff in H_1 being assigned the truth value T. This example works in general.

The first step, therefore, is to describe the salient properties of the set of wff on an open branch of a finished truth tree. This is exactly the goal of the following definition. A (finite) set $\mathscr{H}$ of wff is said to be a *Hintikka set* if it satisfies the following three conditions:

(H1) No atom and its negation belongs to $\mathscr{H}$.

(H2) If an α-wff belongs to $\mathscr{H}$ then both α_1 and α_2 belong to $\mathscr{H}$.

(H3) If a β-wff belongs to $\mathscr{H}$ then at least one of β_1 or β_2 belongs to $\mathscr{H}$.

Hintikka sets are named after the logician Jaakko Hintikka (1929–2015). The key result about Hintikka sets is the following lemma which is nothing other than a generalized version of Example 1.11.12. The proof will be by induction, as in the proof of Lemma 1.3.3, although a slightly different version of induction will be used here called *strong induction*. The proof uses the degree of a wff; you will recall that this is simply the number of connectives in the wff.

Lemma 1.11.13 (Hintikka sets). *Let $\mathscr{H}$ be a Hintikka set. Then there is an assignment of truth values to the atoms in $\mathscr{H}$ that makes every wff in $\mathscr{H}$ true.*

Proof. Let the literals occurring in $\mathscr{H}$ be $l_1, \ldots, l_n$. If $l_i = p_i \in \mathscr{H}$ then assign p_i the truth value T; if $l_i = \neg p_i \in \mathscr{H}$ then assign p_i the truth value F. By (H1), each atom will be assigned at most one truth value. Any other atoms that occur in the wff in $\mathscr{H}$ can be assigned truth values arbitrarily. Call this the *natural truth assignment*. All literals in H now have the truth value T by design. This deals with all wff in the set $\mathscr{H}$ with degree of 0 (and some of degree 1).

Suppose that we have proved that all wff in $\mathscr{H}$ of degree at most n are true under the natural truth assignment. Let $X \in \mathscr{H}$ be a wff of degree $n+1$. We shall prove that it, too, is true under the natural truth assignment. Suppose that X is an α-wff in $\mathscr{H}$. Then by (H2) both α_1 and α_2 belong to $\mathscr{H}$. But, by assumption, α_1 and α_2 are both true under the natural truth assignment. It follows that X will also be true under the natural truth assignment. Suppose that X is a β-wff in $\mathscr{H}$. Then by (H3) at least one of β_1 and β_2 belongs to $\mathscr{H}$. Whichever one it is, if the natural truth assignment makes it true, X will also be true under the natural truth assignment. The proof of the lemma now follows by strong induction. $\qquad\square$

Our second result starts with a wff X and some truth assignment τ making X true. We prove that there is at least one branch in any truth tree with root X which is true under τ. Again, this is best motivated by an example.

Example 1.11.14. We begin with the wff $X = \neg((p \vee q) \to (p \wedge q))$. A truth assignment τ making X true is $p = F$ and $q = T$. The truth tree for X is as follows.

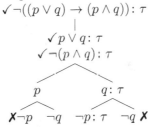

I claim that there is at least one open branch such that all wff on that branch are true under the valuation τ. You can easily check that τ is true for all wff on the second open branch. To illustrate this, I have written the $: \tau$ next to the wff in the second open branch.

$$\checkmark \neg((p \vee q) \to (p \wedge q)): \tau$$

$$\checkmark\, p \vee q : \tau$$
$$\checkmark\, \neg(p \wedge q): \tau$$

$$p \qquad\qquad q: \tau$$

$$\text{✗}\,\neg p \quad \neg q \quad\quad \neg p : \tau \quad \neg q\ \text{✗}$$

Lemma 1.11.15. *Let X be a wff and let τ be a truth assignment to the atoms of X that makes X true. Then for any finished truth tree for X there is at least one open branch such that all wff on that branch are true under τ.*

Proof. Construct a finished truth tree with root X. Write $: \tau$ next to X. I shall describe a process for labelling wff in the truth tree with the decoration $: \tau$ such that if Y is a wff so labelled then $\tau(Y) = T$. When this labelling process is complete we shall show that there will be at least one open branch of the truth tree in which every wff on that branch is labelled $: \tau$; this will prove the lemma. Suppose that there is a path, not a branch, in the truth tree from root to the wff Y such that every wff on that path has been labelled with $: \tau$.

- If Y is an α-formula then both of its successors will be labelled with : τ.

- If Y is a β-formula then at least one of its successors will be labelled with : τ.

Thus every path, not a branch, can be extended. It follows that there must be a branch on which every wff is labelled with : τ. But then it follows that this branch must be open because it cannot contain an atom and its negation since both would be assigned the value T by τ, which is impossible. $\square$

We use Lemmas 1.11.13 and 1.11.15 to prove the main theorem of this section.

Theorem 1.11.16 (Soundness and completeness).

1. **Soundness.** *If X is a tautology, then every finished truth tree with root $\neg X$ is closed.*

2. **Completeness.** *If some finished truth tree with root $\neg X$ is closed, then X is a tautology.*

Proof. (1) Suppose that there is a finished truth tree with root $\neg X$ which is not closed. Then there is an open branch. This open branch is a Hintikka set that contains $\neg X$. By Lemma 1.11.13, there is a truth assignment that makes $\neg X$ true, but this is impossible since $\neg X$ is a contradiction. It follows that there can be no such open branch, and so any finished truth tree with root $\neg X$ is closed.

(2) Suppose that there is a finished truth tree $\mathcal{T}$ with root $\neg X$ which is closed but that X is not a tautology. Then there is some truth assignment τ making $\neg X$ true. Then by Lemma 1.11.15, there is at least one open branch for $\mathcal{T}$ along which τ is true. But this contradicts the fact that $\mathcal{T}$ is closed. It follows that X must be a tautology. $\square$

Exercises 1.11

---

1. Determine whether the following arguments are valid or not using truth trees.

   (a) $p \rightarrow q, r \rightarrow s, p \vee r \models q \vee s$.

   (b) $p \rightarrow q, r \rightarrow s, \neg q \vee \neg s \models \neg p \vee \neg r$.

   (c) $p \rightarrow q, r \rightarrow s, p \vee \neg s \models q \vee \neg r$.

2. Show that the following are tautologies using truth trees.

   (a) $q \rightarrow (p \rightarrow q)$.

(b) $[(p \to q) \land (q \to r)] \to (p \to r)$.

(c) $[(p \to q) \land (p \to r)] \to (p \to (q \land r))$.

(d) $[((p \to r) \land (q \to r)) \land (p \lor q)] \to r$.

3. In [53, Chapter 1], Winnie-the-Pooh makes the following argument.

    (a) There is a buzzing.

    (b) If there is a buzzing then somebody is making the buzzing.

    (c) If somebody is making the buzzing then somebody is bees.

    (d) If somebody is bees then there is honey.

    (e) If there is honey then there is honey for Pooh to eat.

    (f) Therefore there is honey for Pooh to eat.

Using the letters $p, q, r \ldots$ express this argument in symbolic form and use truth trees to determine whether Pooh's argument is valid.

1.12 Sequent calculus

This section is a foray into more advanced territory; it is not needed anywhere else in this book, and so can safely be omitted on a first reading. I have included it for three reasons: first, sequent calculus is important in the further study of logic — so this section is intended as a bridge to that further study; second, the idea of proving statements from axioms by the application of rules is an important one in mathematics and logic, and needed to be touched upon somewhere in this book; third, I wanted to make it clear that truth trees and sequent calculus were two sides of the same coin: in essence (though the devil is in the details, as always) a proof tree in sequent calculus is obtained from a closed truth tree by turning the latter upside down (and then shuffling symbols around).

Notation. In this section, I shall use bold letters, such as $\mathbf{U}$, to mean a set of wff, possibly empty, whereas I shall use ordinary letters, such as X, to mean an individual wff.

Truth trees can be used to *check* whether a wff is a tautology or not whereas in this section, we describe a method for *generating* tautologies called *sequent calculus*. These two methods complement each other or, as mathematicians like to say, the two methods are *dual* to each other.[28] To make our life a

---

[28] There is a similar phenomenon in formal language theory in the duality between acceptors and grammars [43].

little easier, in this section I will only use the connectives $\neg, \wedge, \vee, \rightarrow$ but as we know from Section 1.6 this will not affect the expressiveness of our logic. The method we describe will also involve a change in perspective. First, we shall extend the meaning of the notation introduced in Section 1.10.[29] Let $A_1, \ldots, A_m, B_1, \ldots, B_n$ be wff. We define

$$A_1, \ldots, A_n \models B_1, \ldots, B_n$$

to mean the same thing as

$$A_1, \ldots, A_n \models B_1 \vee \ldots \vee B_n.$$

This has the effect of making the symbol $\models$ symmetrical fore and aft which is mathematically prettier. Second, rather than working with tautologies directly and valid arguments indirectly, I will work with arguments, in our extended sense, directly. This was an approach to mathematical logic developed in [20] by Gerhard Gentzen (1909–1945) who is the presiding genius of this section. Thus given the wff $A_1, \ldots, A_m$ on the left-hand side and the wff $B_1, \ldots, B_n$ on the right-hand side, we want to know whether $A_1, \ldots, A_n \models B_1, \ldots, B_n$ is a valid argument. In order not to pre-empt the answer to this question, we now introduce a new symbol $\Rightarrow$ whose role is simply to act as a punctuation mark between the set of left-hand wff and the set of right-hand wff. More formally, by a *sequent* is meant an expression of the form

$$A_1, \ldots, A_n \Rightarrow B_1, \ldots, B_n$$

where $A_1, \ldots, A_m$ and $B_1, \ldots, B_n$ should each be interpreted as the sets $\{A_1, \ldots, A_m\}$ and $\{B_1, \ldots, B_n\}$, respectively. This means that the order of the wff on the left or of those on the right does not matter and, in addition, repeats on the left or on the right can be introduced or removed as required. Thus sequents will be written in the form $\mathbf{U} \Rightarrow \mathbf{V}$ where both $\mathbf{U}$ and $\mathbf{V}$ are sets of wff. The following definitions are key:

- A sequent $\mathbf{U} \Rightarrow \mathbf{V}$ is said to be *satisfiable* if there is some truth assignment τ that makes the conjunction of the wff in $\mathbf{U}$ true and the disjunction of the wff in $\mathbf{V}$ true.

- A sequent $\mathbf{U} \Rightarrow \mathbf{V}$ is said to be *falsifiable* if there is a truth assignment τ that makes the conjunction of the wff in $\mathbf{U}$ true and the disjunction of the wff in $\mathbf{V}$ false.

- A sequent that cannot be falsified is said to be *valid*.

A sequent of the form $\mathbf{U} \Rightarrow \varnothing$ will simply be written $\mathbf{U} \Rightarrow$. Similarly, a sequent of the form $\varnothing \Rightarrow \mathbf{V}$ will simply be written $\Rightarrow \mathbf{V}$.

---

[29]See also Question 3 of Exercises 1.10.

Observe the following:

- Sequents of the form $\Rightarrow \mathbf{V}$ are valid precisely when the disjunction of the wff in $\mathbf{V}$ is a tautology.

- Sequents of the form $\mathbf{U} \Rightarrow$ are valid precisely when the conjunction of the wff in $\mathbf{U}$ is a contradiction.

- The sequent $\Rightarrow$ is unsatisfiable.

1.12.1 Deduction trees

Our goal is to introduce some rules that will enable us to generate sequents that represent valid arguments. We call this a *proof* of a sequent. The big reveal will be the theorem that says what can be proved is true and what is true can be proved.

Truth is an elusive concept: for the politician, it is infinitely malleable whereas for the scientist, it is the Holy Grail. As we briefly discussed in Section 1.10.2, it was in classical Greece that mathematicians figured out a way of establishing truth: begin with statements known to be true — such statements are called *axioms* — and then find *rules* that are known to preserve truth. Clearly, if you start with axioms and apply the rules a finite number of times then the statements you end up with will all certainly be true; this is called *soundness*. The crucial question is whether *all* truths can be found in this way. If they can, this is called *completeness*. We shall apply this approach to generating valid arguments and prove that we have, indeed, captured all and only the valid ones.

I now define a system I shall call $\mathscr{S}$, where the 'S' stands for 'Smullyan', since this is the system described by him in [67]. Our *axioms* are all sequents of the form

$$\mathbf{U}, X \Rightarrow \mathbf{V}, X.$$

In other words, the set of wff on the left of $\Rightarrow$ should have at least one element in common with the set of wff on the right of $\Rightarrow$. Our *rules* will have one of the following two shapes. The first is

$$\frac{S_1}{S_2}$$

In *proof mode*, this means that given the sequent S_1 as an assumption then the sequent S_2 can be deduced. In *search mode*, this means that S_2 can be deduced from S_1. The second is

$$\frac{S_1 \qquad S_2}{S_3}$$

In *proof mode*, this means that given the sequents S_1 and S_2 as assumptions then the sequent S_3 can be deduced. In *search mode*, this means that S_3 can be deduced from S_1 and S_2.

Thus in proof mode these rules are read from top-to-bottom whereas in search mode they are read from bottom-to-top.

The significance of these two modes will become clearer later. The rules themselves are listed below. Apart from the rule for negation, each rule comes in two flavours: (1) one assumption or (2) two assumptions whereas the rule for negation is either (l) left-moving or (r) right-moving.

Conjunction

$$\frac{\mathbf{U}, X, Y \Rightarrow \mathbf{V}}{\mathbf{U}, X \wedge Y \Rightarrow \mathbf{V}}$$

$$\frac{\mathbf{U} \Rightarrow \mathbf{V}, X \qquad \mathbf{U} \Rightarrow \mathbf{V}, Y}{\mathbf{U} \Rightarrow \mathbf{V}, X \wedge Y}$$

Conjunction (1) simply expresses the fact that the comma on the left-hand side can be interpreted as conjunction.

Disjunction

$$\frac{\mathbf{U} \Rightarrow \mathbf{V}, X, Y}{\mathbf{U} \Rightarrow \mathbf{V}, X \vee Y}$$

$$\frac{\mathbf{U}, X \Rightarrow \mathbf{V} \qquad \mathbf{U}, Y \Rightarrow \mathbf{V}}{\mathbf{U}, X \vee Y \Rightarrow \mathbf{V}}$$

Disjunction (1) simply expresses the fact that the comma on the right-hand side can be interpreted as disjunction.

Negation-creation

$$\frac{\mathbf{U} \Rightarrow \mathbf{V}, X}{\mathbf{U}, \neg X \Rightarrow \mathbf{V}}$$

$$\frac{\mathbf{U}, X \Rightarrow \mathbf{V}}{\mathbf{U} \Rightarrow \mathbf{V}, \neg X}$$

Implication

$$\frac{\mathbf{U}, X \Rightarrow \mathbf{V}, Y}{\mathbf{U} \Rightarrow \mathbf{V}, X \to Y}$$

$$\frac{\mathbf{U} \Rightarrow \mathbf{V}, X \qquad \mathbf{U}, Y \Rightarrow \mathbf{V}}{\mathbf{U}, X \to Y \Rightarrow \mathbf{V}}$$

Implication (1) follows by applying Negation-creation (r) to X and then applying Disjunction (1). Implication (2) follows by applying Negation-creation (l) to X and then applying Disjunction (2).

A *proof* of a sequent is a tree of sequents, I shall call it a *proof tree*, but this time the tree is arboreally correct in having its root at the bottom and its leaves at the top, in such a way that the leaves are axioms and the root is the sequent we are trying to prove, and only the rules above are used in constructing the tree. The axioms, rules and notion of proof we have just described form what is called *sequent calculus*.

Example 1.12.1. The following is a very simple proof in sequent calculus with two leaves only and one application of Implication (2).

$$\frac{p \Rightarrow q, p \qquad p, q \Rightarrow q}{p, p \to q \Rightarrow q}$$

This shows that modus ponens is a valid argument.

> Goal: we want to prove that an argument $\mathbf{U} \vDash \mathbf{V}$ is valid if and only if the sequent $\mathbf{U} \Rightarrow \mathbf{V}$ can be proved in sequent calculus.

Our first result is proved using Question 3 of Exercises 1.10.

Proposition 1.12.2 (Soundness). *If a sequent can be proved then the corresponding argument is valid.*

Before we prove our main theorem, we describe an algorithm. It is here that we shall use our rules in search mode. Given a sequent $\mathbf{U} \Rightarrow \mathbf{V}$, we construct a tree whose root is $\mathbf{U} \Rightarrow \mathbf{V}$ and all of whose vertices are sequents. A leaf of this tree is said to be *finished* if it is either an axiom or all the wff that appear in it are atoms. The tree will be said to be *finished* if all its leaves are finished. The *complexity* of a sequent is the total number of propositional connectives that appear in it. We say that a sequent is *expanded* by a rule of sequent calculus if that sequent appears below the line of a rule. Thus the expansion of a sequent will either result in a new sequent of complexity one less, or two sequents where the sum of their complexities is one less.

The sequent calculus search procedure

Input: A sequent $\mathbf{U} \Rightarrow \mathbf{V}$.

Procedure: The tree is constructed in a *breadth-first* fashion from left-to-right. Each non-finished leaf is expanded; as a strategy, it is best to expand all those leaves first where the rule being applied is non-branching. This process terminates when all leaves are finished.

Output: A finished tree with vertices being sequents whose root is $\mathbf{U} \Rightarrow \mathbf{V}$.

Before continuing, I should explain the meaning of the term *breadth-first*. Suppose you are told that one of the vertices of a tree contained gold. How would you go about searching for it? There are two systematic approaches. The first, called *depth-first*, searches the vertices down to a leaf and then backtracks. The second, called *breadth-first*, searches all the vertices in order, each level at a time. For finite trees either method could be applied but for trees that have infinite branches only breadth-first is applicable. When we discuss FOL, we shall have to consider infinite trees, and so in this book breadth-first search is the natural systematic search procedure.

Any tree obtained by applying the above algorithm to a sequent $\mathbf{U} \Rightarrow \mathbf{V}$ is called a *deduction tree*. It is a *proof tree* when all the leaves are axioms.

Example 1.12.3. To illustrate the above algorithm, we prove the sequent

$$(q \to p) \wedge (p \vee q) \Rightarrow p.$$

We therefore start with this sequent as the root and set about growing the tree above it as part of the proof search. The sequent rules should be used in search mode. Our immediate goal at each step is to simplify the current sequent. The obvious rule to apply is Conjunction (1) (search mode). This yields

$$\frac{(q \to p), (p \vee q) \Rightarrow p}{(q \to p) \wedge (p \vee q) \Rightarrow p}$$

The top sequent is simpler than the bottom one because it has one less connective. At this point we have a choice, but we apply the rule Implication (2) (search mode). This yields

$$\frac{p \vee q \Rightarrow p, q \qquad p, p \vee q \Rightarrow p}{\dfrac{(q \to p), (p \vee q) \Rightarrow p}{(q \to p) \wedge (p \vee q) \Rightarrow p}}$$

We now apply the rule Disjunction (2). This yields

$$\frac{\dfrac{p \Rightarrow p, q \qquad q \Rightarrow p, q}{p \vee q \Rightarrow p, q} \qquad p, p \vee q \Rightarrow p}{\dfrac{(q \to p), (p \vee q) \Rightarrow p}{(q \to p) \wedge (p \vee q) \Rightarrow p}}$$

Next, we consider $p, p \vee q \Rightarrow p$ but this is an axiom. Our search for a proof now ends because all leaves are axioms. Having worked backwards to search for the proof, we can now read the tree forwards from leaves to root and it is this that constitutes the proof.

We can now prove our main result.

Theorem 1.12.4 (Completeness). *Let $A_1, \ldots, A_m \vDash B_1, \ldots, B_n$ be a valid argument. Then the sequent $A_1, \ldots, A_m \Rightarrow B_1, \ldots, B_n$ can be proved.*

Proof. Let $\mathbf{U} \Rightarrow \mathbf{V}$ be a sequent. We prove that if there is a truth assignment τ falsifying the sequent $\mathbf{U} \Rightarrow \mathbf{V}$, then in any deduction tree for $\mathbf{U} \Rightarrow \mathbf{V}$ there is a leaf $\mathbf{X} \Rightarrow \mathbf{Y}$ such that τ falsifies $\mathbf{X} \Rightarrow \mathbf{Y}$, and conversely. The keys to the proof are the following two results. Let τ be a truth assignment. Then τ falsifies S_2 in the rule below

$$\frac{S_1}{S_2}$$

if and only if τ falsifies S_1. Second, τ falsifies S_3 in the rule below

$$\frac{S_1 \qquad S_2}{S_3}$$

if and only if τ falsifies either S_1 or S_2. The proofs of these results are by verification in all cases; see Question 3 of Exercises 1.12.

We can now prove the theorem as stated. Let $A_1, \ldots, A_m \vDash B_1, \ldots, B_n$ be a valid argument. Construct any deduction tree $\mathcal{D}$ for $A_1, \ldots, A_m \Rightarrow B_1, \ldots, B_n$. Suppose that there is a leaf of this tree which is not an axiom. Then it has the form $\mathbf{U} \Rightarrow \mathbf{V}$ where the sets $\mathbf{U}$ and $\mathbf{V}$ have no element in common, by assumption. Define a truth assignment τ as follows: under τ any atoms in $\mathbf{U}$ are assigned the truth value T and any atoms in $\mathbf{V}$ are assigned the truth value F with any remaining atoms assigned truth values arbitrarily. It follows from the first part of the proof that τ falsifies $A_1, \ldots, A_m \Rightarrow B_1, \ldots, B_n$, but this contradicts the fact that $A_1, \ldots, A_m \vDash B_1, \ldots, B_n$ is a valid argument. Thus every leaf in the deduction tree $\mathcal{D}$ is an axiom and so $A_1, \ldots, A_m \Rightarrow B_1, \ldots, B_n$ can be proved. $\square$

Example 1.12.5. The deduction tree for the sequent

$$\Rightarrow (p \to q) \wedge (\neg p \to \neg q)$$

is as follows

$$
\frac{
  \dfrac{q \Rightarrow p}{
    \dfrac{q, \neg p \Rightarrow}{
      \dfrac{\neg p \Rightarrow \neg q}{\Rightarrow \neg p \to \neg q}
    }
  }
  \qquad
  \dfrac{p \Rightarrow q}{\Rightarrow p \to q}
}{\Rightarrow (p \to q) \wedge \neg(p \to \neg q)}
$$

However, neither leaf is an axiom. This means that our sequent cannot be proved. But we can say more. Consider the leaf $p \Rightarrow q$. Assign the value T to p and F to q. This truth assignment makes the wff $(p \to q) \wedge (\neg p \to \neg q)$ take the value F.

1.12.2   Truth trees revisited

In this section, we shall describe the sense in which truth trees and deduction trees are interdefinable. To achieve this goal, we introduce a modification of $\mathscr{S}$, called $\mathscr{S}^*$, and a modification of truth trees, called block truth trees (defined below).[30]

System $\mathscr{S}^*$

The system $\mathscr{S}^*$ consists of sequents of the form $\mathbf{S} \Rightarrow$ called *-sequents. The *-axioms are those sequents of the form $\mathbf{S} \Rightarrow$ where $\mathbf{S}$ contains both an atom and its negation. The *-rules are listed below.

∗-Conjunction

$$\frac{\mathbf{S}, X, Y \Rightarrow}{\mathbf{S}, X \wedge Y \Rightarrow}$$

$$\frac{\mathbf{S}, \neg X \Rightarrow \qquad \mathbf{S}, \neg Y \Rightarrow}{\mathbf{S}, \neg(X \wedge Y) \Rightarrow}$$

---

[30]The system $\mathscr{S}^*$ is obtained from the system $\mathscr{S}$ by moving everything left of the $\Rightarrow$.

---

*-Disjunction

$$\frac{\mathbf{S}, \neg X, \neg Y \Rightarrow}{\mathbf{S}, \neg(X \vee Y) \Rightarrow}$$

$$\frac{\mathbf{S}, X \Rightarrow \qquad \mathbf{S}, Y \Rightarrow}{\mathbf{S}, X \vee Y \Rightarrow}$$

---

*-Negation

$$\frac{\mathbf{S}, X \Rightarrow}{\mathbf{S}, \neg\neg X \Rightarrow}$$

---

*-Implication

$$\frac{\mathbf{S}, X, \neg Y \Rightarrow}{\mathbf{S}, \neg(X \rightarrow Y) \Rightarrow}$$

$$\frac{\mathbf{S}, \neg X \Rightarrow \qquad \mathbf{S}, Y \Rightarrow}{\mathbf{S}, X \rightarrow Y \Rightarrow}$$

Using *-axioms and *-rules we get *-*deduction trees* and *-*proof trees*.

Block truth trees

We now describe the slight modification of truth trees, called *block truth trees*. Rather than α-rules and β-rules, there are $\mathcal{A}$-*rules* and $\mathcal{B}$-*rules*. Let $\mathbf{S}$ be a set of wff. Then the two types of rules are as follows:

- The $\mathcal{A}$-rule:

$$\mathbf{S}, \alpha$$
$$|$$
$$\mathbf{S}, \alpha_1, \alpha_2$$

- The $\mathcal{B}$-rule:

$$\mathbf{S}, \beta$$
$$\overgroup{\mathbf{S}, \beta_1 \quad \mathbf{S}, \beta_2}$$

Block truth trees are constructed and used just like truth trees, except that now a vertex is closed if it contains an atom and its negation. Suppose that a finished block truth tree with root $\neg X$ is closed. Then each leaf is a set of wff that contains an atom and its negation. We now write down explicitly what the $\mathcal{A}$-rules and the $\mathcal{B}$-rules look like.

<table>
<tr><td colspan="4" align="center">$\mathcal{A}$-rules</td></tr>
<tr>
<td>$\mathbf{S}, X \wedge Y$</td>
<td>$\mathbf{S}, \neg(X \vee Y)$</td>
<td>$\mathbf{S}, \neg(X \to Y)$</td>
<td>$\mathbf{S}, \neg\neg X$</td>
</tr>
<tr>
<td>$|$</td>
<td>$|$</td>
<td>$|$</td>
<td>$|$</td>
</tr>
<tr>
<td>$\mathbf{S}, X, Y$</td>
<td>$\mathbf{S}, \neg X, \neg Y$</td>
<td>$\mathbf{S}, X, \neg Y$</td>
<td>$\mathbf{S}, X$</td>
</tr>
</table>

<table>
<tr><td colspan="3" align="center">$\mathcal{B}$-rules</td></tr>
<tr>
<td>$\mathbf{S}, X \vee Y$</td>
<td>$\mathbf{S}, \neg(X \wedge Y)$</td>
<td>$\mathbf{S}, X \to Y$</td>
</tr>
<tr>
<td>$\mathbf{S}, X \quad \mathbf{S}, Y$</td>
<td>$\mathbf{S}, \neg X \quad \mathbf{S}, \neg Y$</td>
<td>$\mathbf{S}, \neg X \quad \mathbf{S}, Y$</td>
</tr>
</table>

In truth trees, truth runs down the branches whereas in block truth trees, truth lives at the leaves.

Example 1.12.6. We construct the block truth tree with root

$$X = \neg((p \vee q) \to (p \wedge q)).$$

It is as follows

$$X$$
$$|$$
$$p \vee q, \neg(p \wedge q)$$

$$p, \neg(p \wedge q) \qquad q, \neg(p \wedge q)$$

$$\mathbf{X} p, \neg p \quad p, \neg q \quad q, \neg p \quad q, \neg q \, \mathbf{X}$$

There is no need to employ checks to show which wff have been used since in block truth trees one is always moving downwards carrying all the information one needs. The price to be paid is that more wff have to be written.

We can now connect truth trees with sequent calculus. We shall do this in three steps. Our first result is not surprising since the rules used in constructing block truth trees only differ from the rules used in the construction of truth trees in the amount of extra luggage one is obliged to carry.

Proposition 1.12.7. *Any finished truth tree can be converted into a finished block truth tree, and vice versa.*

Proof. Let X be wff and let $\mathcal{T}$ be a finished truth tree having X as its root. A corresponding finished block truth tree $\mathcal{T}'$ having X as its root can be constructed in tandom with the construction of the finished truth tree $\mathcal{T}$: when an α-rule is applied in the contruction of $\mathcal{T}$, apply the corresponding $\mathcal{A}$-rule in the construction of $\mathcal{T}'$, and when a β-rule is applied in the contruction of $\mathcal{T}$, apply the corresponding $\mathcal{B}$-rule in the construction of $\mathcal{T}'$. This process can be reversed to construct a finished truth tree $\mathcal{T}$ from a finished block truth tree $\mathcal{T}'$. □

The proof of the next result follows from the observation that the $*$-rules and the $\mathcal{A}$- and $\mathcal{B}$-rules are the same except that the $*$-rules are the $\mathcal{A}$- and $\mathcal{B}$-rules turned upside-down.

Proposition 1.12.8. *Any block truth tree proof can be converted into $*$-proof and vice versa.*

The following example illustrates the effect of applying Proposition 1.12.7 first and then Proposition 1.12.8.

Example 1.12.9. Let

$$X = \neg((p \vee q) \to (p \wedge q)).$$

The following is a truth tree having X as its root:

Next, we convert this truth tree into a block truth tree using the idea of the proof of Proposition 1.12.8:

Finally, we turn the above tree upside down, remove the crosses and insert $\Rightarrow$'s to obtain a $*$-deduction tree for X:

$$
\dfrac{\dfrac{p, \neg p \Rightarrow \qquad p, \neg q \Rightarrow}{p, \neg(p \wedge q) \Rightarrow} \qquad \dfrac{q, \neg p \Rightarrow \qquad q, \neg q \Rightarrow}{q, \neg(p \wedge q) \Rightarrow}}{\dfrac{p \vee q, \neg(p \wedge q) \Rightarrow}{X \Rightarrow}}
$$

Our final result will not be surprising but there is an obstacle to overcome: we obtained ∗-proofs by moving wff in sequent calculus to the left of ⇒ but how on earth do we move right again?

Proposition 1.12.10. *Any ∗-proof can be converted into a proof in sequent calculus, and vice versa.*

Proof. Any proof can be converted into a ∗-proof by simply moving all the wff left across the ⇒ symbol. It remains to show that any ∗-proof can be converted into a proof. Consider the following two deductions:

Negation-annihilation

$$\frac{\mathbf{U} \Rightarrow \mathbf{V}, \neg X}{\mathbf{U}, X \Rightarrow \mathbf{V}}$$

$$\frac{\mathbf{U}, \neg X \Rightarrow \mathbf{V}}{\mathbf{U} \Rightarrow \mathbf{V}, X}$$

It is easy to check that a truth valuation τ falsifies the sequent $\mathbf{U} \Rightarrow \mathbf{V}, X$ if and only if it falsifies the sequent $\mathbf{U}, \neg X \Rightarrow \mathbf{V}$, and it falsifies the sequent $\mathbf{U}, X \Rightarrow \mathbf{V}$ if and only if it falsifies the sequent $\mathbf{U} \Rightarrow \mathbf{V}, \neg X$. Denote by $\mathscr{S}^+$ any proof of a sequent that uses the rules in $\mathscr{S}$ together with the new rule Negation-annihilation. It is clear that this sequent will be valid and so by the Completeness Theorem, Theorem 1.12.4, it can be proved using only the rules in $\mathscr{S}$. □

A more constructive approach to the of key part of the above proof can be found in [7, Lemma 14.15 and Corollary 14.16].

Example 1.12.11. We shall illustrate the relationship between block truth trees and sequent calculus, as described in Proposition 1.12.8 and Proposition 1.12.10, by means of a concrete example. We start with the sequent

$$p \to q, \neg q \Rightarrow \neg p.$$

We move the $\neg p$ to the left and obtain the sequent

$$p \to q, \neg q, \neg\neg p \Rightarrow .$$

Next, we construct the block truth tree that has as its root

$$p \to q, \neg q, \neg\neg p.$$

This is just

$$p \to q, \neg q, \neg\neg p$$
$$|$$
$$p \to q, \neg q, p$$

$$\neg p, \neg q, p \; \textbf{✗} \qquad q, \neg q, p \; \textbf{✗}$$

Now turn this tree upside-down — or right-side up — and write everything as *-sequents:

$$\frac{\neg p, \neg q, p \Rightarrow \qquad q, \neg q, p \Rightarrow}{\dfrac{p \to q, \neg q, p \Rightarrow}{p \to q, \neg q, \neg\neg p \Rightarrow}}$$

This is a *-proof of the sequent $p \to q, \neg q, \neg\neg p \Rightarrow$. We now have to convert this into a proof. Here, we use the Negation annihilation rule:

$$\frac{\dfrac{\neg q, p \Rightarrow p}{\neg p, \neg q, p \Rightarrow} \qquad \dfrac{q, p \Rightarrow q}{q, \neg q, p \Rightarrow}}{\dfrac{\dfrac{p \to q, \neg q, p \Rightarrow}{p \to q, \neg q, \neg\neg p \Rightarrow}}{p \to q, \neg q \Rightarrow \neg p}}$$

Finally, we can also give a proof of the sequent using only the rules $\mathscr{S}$:

$$\frac{\dfrac{\neg q, p \Rightarrow p}{\neg q \Rightarrow \neg p, p} \qquad \dfrac{q \Rightarrow q, \neg p}{\neg q, q \Rightarrow \neg p}}{p \to q, \neg q \Rightarrow \neg p}$$

1.12.3 Gentzen's system LK

We defined sequent calculus using what we called system $\mathscr{S}$, but Gentzen himself used a system he called **LK**.[31] We describe that system here and show that it has exactly the same power as $\mathscr{S}$. Recall that for us, a sequent $\mathbf{U} \Rightarrow \mathbf{V}$ consists of two sets of wff: $\mathbf{U}$ and $\mathbf{V}$. This is convenient for manual calculations but is not as computationally simple as it might appear. In sets, order does not matter and repetitions are ignored. Thus to show, in fact, that two sets are equal, we may actually have to carry out a lot of computation. What Gentzen did was make explicit what those computations should be. First of all, for Gentzen a sequent $\mathbf{a} \Rightarrow \mathbf{b}$ now consisted of two strings, or *sequences*,[32] of symbols, that I shall represent by $\mathbf{a}$ and $\mathbf{b}$. Thus now, order does matter and repeats do make a difference. To regain the flexibility of the set-based system of sequents, three new *structural rules* have to be introduced:

---

[31]From the German *Klassische Prädikatenlogik*.

[32]Whence the name 'sequent calculus'.

Thinning

$$\frac{\mathbf{u} \Rightarrow \mathbf{v}}{\mathbf{u} \Rightarrow \mathbf{v}, X}$$

$$\frac{\mathbf{u} \Rightarrow \mathbf{v}}{\mathbf{u}, X \Rightarrow \mathbf{v}}$$

Contraction

$$\frac{\mathbf{u} \Rightarrow \mathbf{v}, X, X}{\mathbf{u} \Rightarrow \mathbf{v}, X}$$

$$\frac{\mathbf{u}, X, X \Rightarrow \mathbf{v}}{\mathbf{u}, X \Rightarrow \mathbf{v}}$$

Interchange

$$\frac{\mathbf{u} \Rightarrow \mathbf{v}_1, X, Y, \mathbf{v}_2}{\mathbf{u} \Rightarrow \mathbf{v}_1, Y, X, \mathbf{v}_2}$$

$$\frac{\mathbf{u}_1, X, Y, \mathbf{u}_2 \Rightarrow \mathbf{v}}{\mathbf{u}_1, Y, X, \mathbf{u}_2 \Rightarrow \mathbf{v}}$$

In addition, we only need axioms of the form $X \Rightarrow X$. The remaining rules are kept except that sequents are interpreted in terms of strings.

The structural rules enable us to manipulate strings as if they were sets. Consider the string abc. In **LK**, this would be written a, b, c. Thinning certainly enables us to add repeats to obtain, for example, a, a, b, c. Contraction enables us to delete repeats to obtain, say, a, b, c from a, b, b, c. Finally, Interchange enables us to permute the order of the elements of the string to obtain, say, b, a, c from a, b, c. Each string $\mathbf{u}$ gives rise to a unique set $\mathbf{U}$ whose elements are the items that occur in $\mathbf{u}$, whereas each set $\mathbf{U}$ gives rise to many strings $\mathbf{u}$ which are, however, all related by the structural rules.

Theorem 1.12.12. *Let* $\mathbf{u}, \mathbf{v}$ *be strings and let* $\mathbf{U}, \mathbf{V}$ *be the corresponding sets. Then* $\mathbf{u} \Rightarrow \mathbf{v}$ *can be proved in* **LK** *if and only if* $\mathbf{U} \Rightarrow \mathbf{V}$ *can be proved in* $\mathscr{S}$.

Proof. The differences between $\mathscr{S}$ and **LK** really boil down to the differences in the bureaucracy involved. Any sequent proved in the system **LK** can be proved in the system $\mathscr{S}$ by the Completeness Theorem, Theorem 1.12.4. Suppose now that a sequent is proved in the system $\mathscr{S}$. This can be converted into a proof in the system **LK** as follows: any applications of set-theoretic properties are made explicit by means of the structural rules; axioms of the form $\mathbf{U}, X \Rightarrow \mathbf{V}, X$ can be proved in the system **LK** by beginning with the axiom $X \Rightarrow X$ and then repeatedly using thinning, fore and aft. $\square$

Exercises 1.12

---

1. Show that the following arguments are valid using sequent calculus.

    (a) $p \to q, \neg q \models \neg p$.

    (b) $p \to q, q \to r \models p \to r$.

    (c) $p \to q, r \to s, p \vee r \models q \vee s$.

2. Show that the following are tautologies using sequent calculus.

    (a) $p \vee \neg p$.

    (b) $q \to (q \to p)$.

    (c) $((p \to q) \wedge (q \to r)) \to (p \to r)$.

3. Supply the missing proofs of the results stated in the proof of Theorem 1.12.4.

4. Show that the following new rule, called *cut*, is such that if τ is any truth assignment satisfying the two assumptions then it satisfies the conclusion:

$$\frac{\mathbf{U} \Rightarrow \mathbf{V}, X \qquad X, \mathbf{W} \Rightarrow \mathbf{Z}}{\mathbf{U}, \mathbf{W} \Rightarrow \mathbf{V}, \mathbf{Z}}$$

The cut rule is important in the more advanced theory of sequent calculus.

Chapter 2

Boolean algebras

"He's out" said Pooh sadly, "That's what it is. He's not in." —
Winnie-the-Pooh.

Unlike the other two chapters, this one is not about logic per se but about
an algebraic system that arises from logic. The following table shows how symbols of propositional logic are to be replaced by symbols of Boolean algebra,
although mathematically there is more to it than just using different notation:

Propositional logic	Boolean algebra
$\equiv$	$=$
$\vee$	$+$
$\wedge$	$\cdot$
$\neg$	$^-$
t	1
f	0

Boolean algebra provides a bridge between logic and the real world because
it is used in the design of the kinds of circuits needed to build computers.
You can, if you wish, skip this chapter or return to it after Chapter 3; only
Section 2.1 is needed for Chapter 3.

2.1 More set theory

Sets were briefly touched upon in Section 1.2, but now we need to study
them in more detail. We can sometimes define infinite sets by using curly
brackets but then, because we cannot list all elements in an infinite set, we
use '...' to mean 'and so on in the obvious way'. This can also be used to define
big finite sets where there is an obvious pattern. However, the most common
way of describing a set is to say what properties an element must have to
belong to it — in other words, what the membership conditions of the set are.
By a *property* or *1-place predicate* we mean a sentence containing a variable
such as x so that the sentence becomes true or false depending on what we

substitute for x. We shall say much more about properties in Chapter 3 since they are an important ingredient in first-order logic. For example, the sentence 'x is an even natural number' is true when x is replaced by 2 and false when x is replaced by 3. If we abbreviate 'x is an even natural number' by $E(x)$ then the set of even natural numbers is the set of all natural numbers n such that $E(n)$ is true. This set is written $\{x\colon E(x)\}$ or $\{x \mid E(x)\}$. More generally, if $P(x)$ is any property then $\{x\colon P(x)\}$ means 'the set of all things x that satisfy the condition P'. The following notation will be useful when we come to study first-order logic.

Examples 2.1.1.

1. The set $\mathbb{N} = \{0, 1, 2, 3, \ldots\}$ of all *natural numbers*. Caution is required here since some books eccentrically do not regard 0 as a natural number.

2. The set $\mathbb{Z} = \{\ldots, -3, -2, -1, 0, 1, 2, 3, \ldots\}$ of all *integers*. The reason $\mathbb{Z}$ is used to designate this set is because '$\mathbb{Z}$' is the first letter of the word 'Zahl', the German word for number.

3. The set $\mathbb{Q}$ of all *rational numbers*. That is, those numbers that can be written as quotients (whence the '$\mathbb{Q}$') of integers with non-zero denominators.

4. The set $\mathbb{R}$ of all *real numbers*. That is, all numbers which can be represented by decimals with potentially infinitely many digits after the decimal point.

Given a set A, a new set B can be formed by *choosing* elements from A to put into B. We say that B is a *subset* of A, denoted by $B \subseteq A$. In mathematics, the word 'choose', unlike in polite society, also includes the possibility of *choosing nothing* and the possibility of *choosing everything*. In addition, there does not need to be any rhyme or reason to your choices: you can pick elements 'at random' if you want. If $A \subseteq B$ and $A \neq B$ then we say that A is a *proper subset* of B.

Examples 2.1.2.

1. $\varnothing \subseteq A$ for every set A, where we choose no elements from A.

2. $A \subseteq A$ for every set A, where we choose all the elements from A.

3. $\mathbb{N} \subseteq \mathbb{Z} \subseteq \mathbb{Q} \subseteq \mathbb{R}$. Observe that $\mathbb{Z} \subseteq \mathbb{Q}$ because an integer n is equal to the rational number $\frac{n}{1}$.

4. $\mathbb{E}$, the set of even natural numbers, is a subset of $\mathbb{N}$.

5. $\mathbb{O}$, the set of odd natural numbers, is a subset of $\mathbb{N}$.

Example 2.1.3. The idea that sets are defined by properties is natural but logically suspect. It seems obvious that given a property $P(x)$, there is a corresponding set $\{x\colon P(x)\}$ of all those things that have that property. But we now describe a result famous in the history of mathematics called *Russell's paradox*, named after Bertrand Russell (1872–1970), which shows that just because something is obvious does not make it true. Define $\mathscr{R} = \{x\colon x \notin x\}$. In other words: the set of all sets that do not contain themselves as an element. For example, $\varnothing \in \mathscr{R}$. We now ask the question: is $\mathscr{R} \in \mathscr{R}$? There are only two possible answers and we investigate them both.

1. Suppose that $\mathscr{R} \in \mathscr{R}$. This means that $\mathscr{R}$ must satisfy the entry requirements to belong to $\mathscr{R}$ which it can only do if $\mathscr{R} \notin \mathscr{R}$.

2. Suppose that $\mathscr{R} \notin \mathscr{R}$. This means that $\mathscr{R}$ does not satisfy the entry requirements to belong to $\mathscr{R}$ which it can only do if $\mathscr{R} \in \mathscr{R}$.

Thus exactly one of $\mathscr{R} \in \mathscr{R}$ and $\neg(\mathscr{R} \in \mathscr{R})$ must be true but assuming one implies the other. Our only way out is to conclude that, whatever $\mathscr{R}$ might be, it is not a set. This contradicts the 'obvious statement' we began with.[1] However disconcerting you might find this, imagine how much more so it was for the mathematician Gottlob Frege (1848–1925). He was working on a book which based the development of mathematics on sets when he received a letter from Russell describing this paradox thereby undermining what Frege was attempting to achieve.

The example above is a warning on the perils of trying to talk about all sets. In practice, we are usually interested in studying sets that are subsets of some, suitably big, set. Grandiosely, such a set might be called a *universe*. I shall return to this idea of universes later.

We now define three operations on sets that are based on the PL connectives $\wedge$, $\vee$ and $\neg$. They are called *Boolean operations*, named after George Boole. Let A and B be sets.

Define a set, called the *intersection* of A and B, denoted by $A \cap B$, whose elements consist of all those elements that belong to A **and** B.

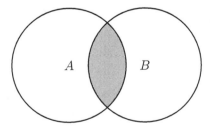

More formally, we can write

$$A \cap B = \{x\colon (x \in A) \wedge (x \in B)\}.$$

---

[1]If you want to understand how to escape this predicament, you will have to study *axiomatic set theory*.

Define a set, called the *union* of A and B, denoted by $A \cup B$, whose elements consist of all those elements that belong to A **or** B.

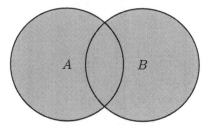

More formally, we can write

$$A \cup B = \{x \colon (x \in A) \vee (x \in B)\}.$$

Define a set, called the *set difference* or *relative complement* of A and B, denoted by $A \setminus B$,[2] whose elements consist of all those elements that belong to A **and not** to B.

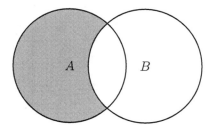

More formally, we can write

$$A \setminus B = \{x \colon (x \in A) \wedge \neg(x \in B)\}.$$

The diagrams used to illustrate the above definitions are called *Venn diagrams*, after John Venn (1834–1923), where a set is represented by a region in the plane. Sets A and B are said to be *disjoint* if $A \cap B = \varnothing$.

Example 2.1.4. Recall that $\mathbb{N}$ is the set of natural numbers, $\mathbb{E}$ is the set of even numbers and $\mathbb{O}$ is the set of odd numbers. Then $\mathbb{N} = \mathbb{E} \cup \mathbb{O}$ since each natural number is either even or odd. However, $\mathbb{E} \cap \mathbb{O} = \varnothing$ since no natural number is both odd and even. It follows that the sets $\mathbb{E}$ and $\mathbb{O}$ are disjoint. Thus each natural number is either odd or even but not both (so, 'exclusive or'). Observe that $\mathbb{N} \setminus \mathbb{E} = \mathbb{O}$ and $\mathbb{N} \setminus \mathbb{O} = \mathbb{E}$.

Theorem 2.1.5 (Properties of Boolean operations). *Let A, B and C be any sets.*

1. $A \cap (B \cap C) = (A \cap B) \cap C$. *Intersection is associative.*

---

[2]Sometimes denoted by $A - B$.

2. $A \cap B = B \cap A$. *Intersection is commutative.*

3. $A \cap \varnothing = \varnothing = \varnothing \cap A$. *The empty set is the zero for intersection.*

4. $A \cup (B \cup C) = (A \cup B) \cup C$. *Union is associative.*

5. $A \cup B = B \cup A$. *Union is commutative.*

6. $A \cup \varnothing = A = \varnothing \cup A$. *The empty set is the identity for union.*

7. $A \cap (B \cup C) = (A \cap B) \cup (A \cap C)$. *Intersection distributes over union.*

8. $A \cup (B \cap C) = (A \cup B) \cap (A \cup C)$. *Union distributes over intersection.*

9. $A \setminus (B \cup C) = (A \setminus B) \cap (A \setminus C)$. *De Morgan's law, part one.*

10. $A \setminus (B \cap C) = (A \setminus B) \cup (A \setminus C)$. *De Morgan's law, part two.*

11. $A \cap A = A$. *Intersection is idempotent.*

12. $A \cup A = A$. *Union is idempotent.*

Proof. I shall just look at property (7) as an example of what can be done. To *illustrate* this property, we can use Venn diagrams. The Venn diagram for

$$(A \cap B) \cup (A \cap C)$$

is given below:

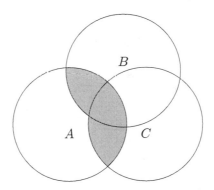

This is exactly the same as the Venn diagram for

$$A \cap (B \cup C).$$

It follows that

$$A \cap (B \cup C) = (A \cap B) \cup (A \cap C)$$

at least as far as Venn diagrams are concerned.

To *prove* property (7), we have to proceed more formally and use PL. We use the fact that

$$p \wedge (q \vee r) \equiv (p \wedge q) \vee (p \wedge r).$$

Our goal is to prove that

$$A \cap (B \cup C) = (A \cap B) \cup (A \cap C).$$

To do this, we have to prove that the set of elements belonging to the left-hand side is the same as the set of elements belonging to the right-hand side. An element x either belongs to A or it does not. Similarly, it either belongs to B or it does not, and it either belongs to C or it does not. Define p to be the statement '$x \in A$'. Define q to be the statement '$x \in B$'. Define r to be the statement '$x \in C$'. If p is true then x is an element of A, and if p is false then x is not an element of A. Using now the definitions of the Boolean operations, it follows that $x \in A \cap (B \cup C)$ precisely when the statement $p \wedge (q \vee r)$ is true. Similarly, $x \in (A \cap B) \cup (A \cap C)$ precisely when the statement $(p \wedge q) \vee (p \wedge r)$ is true. But these two statements have the same truth tables. It follows that an element belongs to the left-hand side precisely when it belongs to the right-hand side. Consequently, the two sets are equal. □

The fact that $A \cap (B \cap C) = (A \cap B) \cap C$ means that we can just write $A \cap B \cap C$ unambiguously without brackets. Similarly, we can write $A \cup B \cup C$ unambiguously. This can be extended to any number of unions and any number of intersections; see Section 1.4.2.

Example 2.1.6. The notion of subset enables us to justify an important method of proof, called proof by induction, that we used in the proof of Lemma 1.3.3 and Lemma 1.11.13 and which we shall use again in the proof of Proposition 3.2.7. To do this, I will need the following property of the natural numbers: every non-empty subset has a smallest element. Proof by induction is used in the following situation: we wish to prove an infinite number of statements $p_0, p_1, p_2, \ldots$. This seems to be impossible but there are many instances where not only is it possible, it is routine. Define X to be the subset of $\mathbb{N}$ of those i where p_i is known to be true. We make two assumptions about X:

1. $0 \in X$.

2. If $N \in X$ then $N + 1 \in X$.

Assumption (1) says that the statement p_0 is true. Assumption (2) says that if the statement p_N is true then the statement p_{N+1} must be true as well. I claim that if these two assumptions hold then $X = \mathbb{N}$ which proves that p_i is true for all $i \in \mathbb{N}$. I now prove the claim. Suppose that $X \neq \mathbb{N}$. Then the set $\mathbb{N} \setminus X$ is non-empty. It therefore has a smallest element. Call it m. By assumption, $m - 1 \in X$. But by property (2), if $m - 1 \in X$ then $m \in X$, which is a contradiction. It follows that $X = \mathbb{N}$, as required.

This result justifies proof by induction. Here is an example. Let p_n be the statement '$3^n - 1$ is even' where $n \in \mathbb{N}$. When $n = 0$ we have that $3^0 - 1 = 0$ which is even. Thus assumption (1) above holds. We now verify that assumption (2) is true. Assume that $3^N - 1$ is even. We prove that $3^{N+1} - 1$ is even. To do this requires a little mathematical manipulation. Observe that

$$3^{N+1} - 1 = 3 \cdot 3^N - 1 = 2 \cdot 3^N + \left(3^N - 1\right).$$

But, by assumption, $3^N - 1$ is even and the sum of two even numbers is even, consequently $3^{N+1} - 1$ is also even.

This basic notion of induction can be extended. Let $m \geq 1$ be a natural number. Define

$$\mathbb{N}^{\geq m} = \mathbb{N} \setminus \{0, \ldots, m - 1\}.$$

Let $X \subseteq \mathbb{N}^{\geq m}$. We suppose that X satisfies assumption (2) and a modified version of assumption (1): namely, $m \in X$. Then by a similar argument, we can show that $X = \mathbb{N}^{\geq m}$. This modified form of induction can be used to prove all the statements $p_m, p_{m+1}, p_{m+2}, \ldots$.

Here is an example. Define $0! = 1$ and $n! = n(n - 1)!$ when $n \geq 1$. Let p_n be the statement '$n! > 2^n$'. We prove that p_n is true for all natural numbers $n \geq 4$. (You can check that the result is *not* true for $n = 0, 1, 2, 3$.) When $n = 4$ we have that $n! = 24$ and $2^4 = 16$. Thus p_4 is true. We now prove that if p_N is true so too is p_{N+1}. Observe that

$$(N + 1)! = (N + 1)N! > (N + 1)2^N > (1 + 1)2^N = 2^{N+1}$$

where we have used the fact that $N > 1$.

Finally, there is a version of induction known as *strong induction*. This is based on the following result which is left as an exercise to prove since the proof is similar to the ones we gave above. Let $X \subseteq \mathbb{N}$ be a subset satisfying the following two assumptions:

1. $0 \in X$.

2. If $\{0, 1, \ldots, N\} \subseteq \mathbb{N}$ then $\{0, 1, \ldots, N + 1\} \subseteq \mathbb{N}$.

Then $X = \mathbb{N}$. As with induction, strong induction also applies to the set $\mathbb{N}^{\geq m}$.

Here is an example of strong induction. Let p_n be the statement 'n can be written as a product of primes (or is prime itself)' where $n \geq 2$. Now p_2 is true since 2 is a prime. Suppose that all the statements $p_2, \ldots, p_N$ are true. Consider now the number $N + 1$. There are two possibilities: it is either prime or not prime. If $N + 1$ is prime then the statement p_{N+1} is true. If $N + 1$ is not prime then it can written as a product $N + 1 = mn$ where $2 \leq m, n \leq N$. But, by assumption, the statements p_m and p_n are both true and so each of m and n can be written as products of primes. It follows that $N + 1$ can be written as a product of primes and so p_{N+1} is true.

Although the theory behind proof by induction is simple, actually using it requires mathematical insight.

Exercises 2.1

1. Let $X = \{1, 2, 3, 4, 5, 6, 7, 8, 9, 10\}$. Write down the following subsets of X:

   (a) The subset A of even elements of X.

   (b) The subset B of odd elements of X.

   (c) $C = \{x \colon (x \in X) \wedge (x \geq 6)\}$.

   (d) $D = \{x \colon (x \in X) \wedge (x > 10)\}$.

   (e) $E = \{x \colon (x \in X) \wedge (x \text{ is prime})\}$.

   (f) $F = \{x \colon (x \in X) \wedge ((x \leq 4) \vee (x \geq 7))\}$.

2. Let A and B be sets. Define the operation $\triangle$ by

$$A \triangle B = (A \setminus B) \cup (B \setminus A).$$

   Prove that $(A \triangle B) \triangle C = A \triangle (B \triangle C)$ for any sets A, B, C.

3. Prove that $(A \setminus B) \setminus C = A \setminus (B \cup C)$.

4. Prove that $A \setminus (B \setminus C) = (A \setminus B) \cup (A \cap C)$.

2.2 Boolean algebras

In 1854, George Boole wrote a book [6] where he tried to show that some of the ways in which we reason could be studied mathematically. This led, perhaps indirectly [25], to what are now called Boolean algebras and was also important in the development of propositional logic and set theory.

2.2.1 Motivation

To motivate the definition of Boolean algebras, we return to the set operations defined in the previous section but this time we shall only look at subsets of some fixed set X (what we called a universe, earlier). The set whose elements are all the subsets of X is called the *power set* of X and is denoted by $\mathsf{P}(X)$. It is important not to forget that the power set of a set X contains both $\varnothing$ and X as elements.

Example 2.2.1. We find all the subsets of the set $X = \{a, b, c\}$ and so determine the power set of X. First there is the subset with no elements, the empty set. Then there are the subsets that contain exactly one element: $\{a\}, \{b\}, \{c\}$. Then there are the subsets containing exactly two elements: $\{a, b\}, \{a, c\}, \{b, c\}$. Finally, there is the whole set, X. It follows that X has eight subsets and so

$$\mathsf{P}(X) = \{\varnothing, \{a\}, \{b\}, \{c\}, \{a, b\}, \{a, c\}, \{b, c\}, X\}.$$

With the set X fixed in the background, we can now define a new unary operation. Let $A \subseteq X$. Define $\overline{A} = X \setminus A$. This is called the *complement* of A.

Example 2.2.2. Let $X = \{a, b, c\}$. Then $\overline{\{a\}} = \{b, c\}$ and $\overline{\{a, b, c\}} = \varnothing$.

We now specialize Theorem 2.1.5 to the case where the sets A, B, C are chosen to be subsets of some fixed set X.

Theorem 2.2.3 (The power set is a Boolean algebra). *Let X be a fixed set and let A, B and C be any subsets.*

1. $A \cup (B \cup C) = (A \cup B) \cup C$.

2. $A \cup B = B \cup A$.

3. $A \cup \varnothing = A$.

4. $A \cap (B \cap C) = (A \cap B) \cap C$.

5. $A \cap B = B \cap A$.

6. $A \cap X = A$.

7. $A \cap (B \cup C) = (A \cap B) \cup (A \cap C)$.

8. $A \cup (B \cap C) = (A \cup B) \cap (A \cup C)$.

9. $A \cup \overline{A} = X$.

10. $A \cap \overline{A} = \varnothing$.

At this point, you should compare the above equations with the following logical equivalences proved in Exercises 1.4; it is here that the logical constants t and f come into their own.

1. $(p \vee q) \vee r \equiv p \vee (q \vee r)$.

2. $p \vee q \equiv q \vee p$.

3. $p \vee f \equiv p$.

4. $(p \wedge q) \wedge r \equiv p \wedge (q \wedge r)$.

5. $p \wedge q \equiv q \wedge p$.

6. $p \wedge t \equiv p$.

7. $p \wedge (q \vee r) \equiv (p \wedge q) \vee (p \wedge r)$.

8. $p \vee (q \wedge r) \equiv (p \vee q) \wedge (p \vee r)$.

9. $p \vee \neg p \equiv t$.

10. $p \wedge \neg p \equiv f$.

Each of these logical equivalences can be used to prove the corresponding result in Theorem 2.2.3. The concept of a Boolean algebra is obtained by abstracting the properties listed in Theorem 2.2.3. A Boolean algebra is defined by equations and so is an *algebraic system* rather than a *logical system*. It is adapted to dealing with truth tables and so with circuits, as we shall see.

2.2.2   Definition and examples

Formally, a *Boolean algebra* is defined by the following data $(B, +, \cdot, ^-, 0, 1)$ where: B is a set that carries the structure; $+$ and $\cdot$ are binary operations, meaning that they have two ordered inputs and one output; $a \mapsto \bar{a}$ is a unary operation, meaning that it has one input and one output, and there are two special elements of B, 0 and 1. In addition, the following ten axioms are required to hold:

(B1) $(x + y) + z = x + (y + z)$.

(B2) $x + y = y + x$.

(B3) $x + 0 = x$.

(B4) $(x \cdot y) \cdot z = x \cdot (y \cdot z)$.

(B5) $x \cdot y = y \cdot x$.

(B6) $x \cdot 1 = x$.

(B7) $x \cdot (y + z) = x \cdot y + x \cdot z$.

(B8) $x + (y \cdot z) = (x + y) \cdot (x + z)$.

(B9) $x + \bar{x} = 1$.

(B10) $x \cdot \bar{x} = 0$.

I shall call the operation $\cdot$ *multiplication*, and usually write ab rather than $a \cdot b$, and I shall call the operation $+$ *addition*.

These axioms are organized as follows. The first group of three, (B1), (B2) and (B3), deals with the properties of $+$ on its own: brackets, order, special element. The second group of three, (B4), (B5) and (B6), deals with the properties of $\cdot$ on its own: brackets, order, special element. The third group, (B7) and (B8), deals with how $+$ and $\cdot$ interact, with axiom (B8) being decidedly odd looking from the perspective of high-school algebra. The final group, (B9) and (B10), deals with the properties of $a \mapsto \bar{a}$, called *complementation*. Because of (B1) and (B4), we can write sums and products without the need for brackets. We shall now describe three examples of Boolean algebras with, ironically, the last and simplest being the most important when it comes to circuit design.

Example 2.2.4. *The Lindenbaum algebra.* This is the Boolean algebra associated with PL. It can be described, in a rather cavalier fashion, as being obtained from PL by replacing $\equiv$ with $=$. How to do this in a mathematically legitimate way is quite involved so I shall only sketch out the idea here. Denote by **Prop** the set of all wff in propositional logic. The key to the construction of a Boolean algebra from **Prop** is the behaviour of the relation $\equiv$ as described by Proposition 1.4.13. The first three properties there tell us that $\equiv$ is an *equivalence relation*; I shall not define what this means but I shall describe what it implies. For each wff A, we define the set

$$[A] = \{B : B \in \textbf{Prop} \text{ and } B \equiv A\}$$

of all wff logically equivalent to A. These sets have an interesting property by virtue of the fact that $\equiv$ is an equivalence relation: either $[C] \cap [D] = \varnothing$ or $[C] = [D]$. Define $\mathbb{L}$ to be the set of all these sets. Thus

$$\mathbb{L} = \{[A] : A \in \textbf{Prop}\}.$$

We now make the following definitions

$$[C] \cdot [D] = [C \wedge D], \quad [C] + [D] = [C \vee D] \text{ and } \overline{[C]} = [\neg C].$$

These definitions really do make sense; for example, if $[C] = [C']$ and $[D] = [D']$ then $[C \vee D] = [C' \vee D']$. Finally, define $1 = [\boldsymbol{t}]$ and $0 = [\boldsymbol{f}]$. Using Proposition 1.4.13, it is then possible to show (with some work) that with these definitions, $\mathbb{L}$ really is a Boolean algebra. This algebra is named after Adolf Lindenbaum (1904–1941?)[3] and finds an application in more advanced work to prove results about PL in an algebraic way [10, Chapter 2], [71, Chapter 6].

---

[3] Lindenbaum was arrested by the Gestapo in 1941 and then, at some later time, murdered by the Nazis. This accounts for the uncertainty over the exact year in which he died. I would like to thank the copy-editor, Cindy Gallardo, for drawing my attention to this. In the preface to his book [72], Tarski pays tribute to Lindenbaum as a former student and colleague, and as one engulfed by the 'wave of organized totalitarian barbarism'.

Example 2.2.5. *The power set Boolean algebra.* This is an important class of examples of Boolean algebras which arise naturally from set theory. Let X be a set and recall that $P(X)$ is the set of all subsets of X. We now define the following operations on $P(X)$.

Boolean algebra	Power set
$+$	$\cup$
$\cdot$	$\cap$
$^-$	$^-$
1	X
0	$\varnothing$

By Theorem 2.2.3, it follows that

$$(P(X), \cup, \cap, ^-, \varnothing, X)$$

is a Boolean algebra. In fact, it can be proved that the finite Boolean algebras always look like power set Boolean algebras.

Example 2.2.6. *The two-element Boolean algebra* $\mathbb{B}$. This is the Boolean algebra we shall use in circuit design. It is defined as follows. Put $\mathbb{B} = \{0, 1\}$. We define operations $^-$, $\cdot$ and $+$ by means of the following tables. They are the same as the truth table for $\neg$, $\wedge$ and $\vee$ except that we replace T by 1 and F by 0.

x	$\bar{x}$
1	0
0	1

x	y	$x \cdot y$
1	1	1
1	0	0
0	1	0
0	0	0

x	y	$x + y$
1	1	1
1	0	1
0	1	1
0	0	0

You can check that $\mathbb{B}$ is essentially the same as the Boolean algebra $P(\{a\})$.

2.2.3 Algebra in a Boolean algebra

Boolean algebra is similar to, but also different from, the kind of algebra you learnt at school. Just how different is illustrated by the following results that will be true in all Boolean algebras because they are proved from the Boolean algebra axioms.

Proposition 2.2.7. *Let B be a Boolean algebra and let $a, b \in B$.*

1. $a^2 = a \cdot a = a$. *Idempotence.*

2. $a + a = a$. *Idempotence.*

3. $a \cdot 0 = 0$. *The element 0 is the zero for multiplication.*

4. $1 + a = 1$. *The element 1 is the zero for addition.*

5. $a = a + a \cdot b$. *Absorption law.*

6. $a + b = a + \bar{a} \cdot b = a + b \cdot \bar{a}$. *Difference law.*

Proof. (1)

$$
\begin{aligned}
a &= a \cdot 1 \text{ by (B6)} \\
&= a \cdot (a + \bar{a}) \text{ by (B9)} \\
&= a \cdot a + a \cdot \bar{a} \text{ by (B7)} \\
&= a^2 + 0 \text{ by (B10)} \\
&= a^2 \text{ by (B3)}.
\end{aligned}
$$

(2) is proved in a very similar way to (1). Use the fact that the Boolean algebra axioms come in pairs where $\cdot$ and $+$ are interchanged and 0 and 1 are interchanged. This is an aspect of what is called *duality*.

(3)

$$
\begin{aligned}
a \cdot 0 &= a \cdot (a \cdot \bar{a}) \text{ by (B10)} \\
&= (a \cdot a) \cdot \bar{a} \text{ by (B4)} \\
&= a \cdot \bar{a} \text{ by (1) above} \\
&= 0 \text{ by (B10)}.
\end{aligned}
$$

(4) The dual proof to (3).

(5)

$$
\begin{aligned}
a + a \cdot b &= a \cdot 1 + a \cdot b \text{ by (B6)} \\
&= a \cdot (1 + b) \text{ by (B7)} \\
&= a \cdot 1 \text{ by (4) above} \\
&= a \text{ by (B6)}.
\end{aligned}
$$

(6)

$$
\begin{aligned}
a + b &= a + 1b \text{ by (B6)} \\
&= a + (a + \bar{a})b \text{ by (B9)} \\
&= a + ab + \bar{a}b \text{ by (B7) and (B5)} \\
&= a + \bar{a}b \text{ by (5) above}.
\end{aligned}
$$

In the above proofs, the dual case is left as an exercise. □

These properties can be used to simplify Boolean expressions.

Example 2.2.8. We shall simplify

$$x + yz + \bar{x}y + x\bar{y}z.$$

At each stage in our argument, we will be explicit about what properties we are using to justify our claims. I shall use associativity of addition throughout to avoid too many brackets. We have that

$$
\begin{aligned}
x + yz + \bar{x}y + x\bar{y}z &= (x + \bar{x}y) + yz + x\bar{y}z \text{ by commutativity} \\
&= (x + y) + yz + x\bar{y}z \text{ by absorption} \\
&= x + (y + yz) + x\bar{y}z \text{ by associativity} \\
&= x + y + x\bar{y}z \text{ by absorption} \\
&= x + (y + \bar{y}(xz)) \text{ by commutativity and associativity} \\
&= x + (y + xz) \text{ by difference law} \\
&= (x + xz) + y \text{ by commutativity and associativity} \\
&= x + y \text{ by absorption.}
\end{aligned}
$$

We may also use Venn diagrams to illustrate certain Boolean expressions and so help us in simplifying them.

Example 2.2.9. We return to the Boolean expression

$$x + yz + \bar{x}y + x\bar{y}z$$

that we simplified above. We interpret this in set theory where the Boolean variables x, y, z are replaced by sets X, Y, Z. Observe that $\bar{x}y = y\bar{x}$ translates into $Y \setminus X$. Thus the above Boolean expression is the set

$$X \cup (Y \cap Z) \cup (Y \setminus X) \cup ((X \cap Z) \setminus Y).$$

If you draw the Venn diagram of this set you get exactly $X \cup Y$. This translates into the Boolean expression $x + y$ which is what we obtained when we simplified the Boolean expression.

A more complex example of proving results about Boolean algebras is included below. It is important to know the result, if not the proof. It shows that the analogue of De Morgan's laws hold in any Boolean algebra.

Proposition 2.2.10. *Let B be a Boolean algebra.*

1. *If $a + b = 1$ and $ab = 0$ then $b = \bar{a}$.*

2. *$\bar{\bar{a}} = a$.*

3. *$\overline{(a + b)} = \bar{a}\bar{b}$.*

4. *$\overline{ab} = \bar{a} + \bar{b}$.*

Proof. (1) We are given that $a + b = 1$ and $ab = 0$, and our goal is to show that $b = \bar{a}$. The path from the former to the latter is not obvious.

We know that $a + \bar{a} = 1$ by (B9). Thus $a + b = a + \bar{a}$. Multiply both sides of the above equation on the left by b to get $b(a + b) = b(a + \bar{a})$. This gives $ba + b^2 = ba + b\bar{a}$ by (B7). Now $ba = ab$ by the commutativity of multiplication and $b^2 = b$. It follows that $ab + b = ab + \bar{a}b$. But $ab = 0$, by assumption. Thus

$$b = \bar{a}b.$$

We shall now prove that $\bar{a} = \bar{a}b$, which will prove our claim. We are given that $a + b = 1$. Multiply both sides of this equation on the left by $\bar{a}$ to get $\bar{a}(a + b) = \bar{a}1 = \bar{a}$ by (B6). By (B7), we have that $\bar{a}a + \bar{a}b = \bar{a}$. By (B10), we have that $\bar{a}a = 0$. It follows that

$$\bar{a} = \bar{a}b.$$

We have therefore proved that $b = \bar{a}$.

(2) By (B9), we have that $\bar{a} + \bar{\bar{a}} = 1$, and by (B10), we have that $\bar{a}\bar{\bar{a}} = 0$. We also have by (B9), (B10) and commutativity that $\bar{a} + a = 1$ and $\bar{a}a = 0$. It follows that $\bar{\bar{a}} = a$ by part (1).

(3) By (B9) and (B10), we have that

$$(a + b) + \overline{(a + b)} = 1 \text{ and } (a + b)\overline{(a + b)} = 0.$$

We now calculate

$$(a + b) + \bar{a}\bar{b} = (a + \bar{a}\bar{b}) + b = a + (\bar{b} + b) = a + 1 = 1$$

and

$$(a + b)\bar{a}\bar{b} = a\bar{a}\bar{b} + b\bar{a}\bar{b} = 0 + 0 = 0.$$

It follows by part (1) that $\overline{(a + b)} = \bar{a}\bar{b}$.

(4) The proof of this case is similar to that in part (3). □

Exercises 2.2

1. This question concerns the following diagram where $A, B, C \subseteq X$. Use the Boolean operations of intersection, union and complementation to describe each of the eight numbered regions. For example, region (1) is $A \cap B \cap C$.

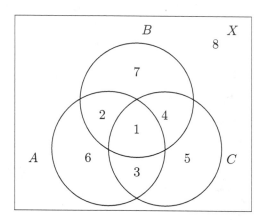

2. Find all subsets of the set $\{a, b, c, d\}$.

3. Let $A, B, C \subseteq X$. By drawing Venn diagrams, illustrate the following set equalities.

    (a) $A \cup (B \cap C) \cup (\overline{A} \cap B) \cup (A \cap \overline{B} \cap C) = A \cup B$.
    (b) $(\overline{A} \cap \overline{B} \cap C) \cup (\overline{A} \cap B \cap C) \cup (A \cap \overline{B}) = (A \cap \overline{B}) \cup (\overline{A} \cap C)$.
    (c) $(A \cap B \cap C) \cup (\overline{A} \cap B \cap C) \cup \overline{B} \cup \overline{C} = X$.

4. Prove the following.

    (a) $a \vee (b \wedge c) \vee (\neg a \wedge b) \vee (a \wedge \neg b \wedge c) \equiv a \vee b$.
    (b) $(\neg a \wedge \neg b \wedge c) \vee (\neg a \wedge b \wedge c) \vee (a \wedge \neg b) \equiv (a \wedge \neg b) \vee (\neg a \wedge c)$.
    (c) $\vDash (a \wedge b \wedge c) \vee (\neg a \wedge b \wedge c) \vee \neg b \vee \neg c$.

5. What is the connection between Questions 3 and 4 above?

6. Prove that $a + a = a$ for all $a \in B$ in a Boolean algebra.

7. Prove that $0a = 0$ for all $a \in B$ in a Boolean algebra.

8. Prove that $1 + a = 1$ for all $a \in B$ in a Boolean algebra.

9. Prove the following laws in any Boolean algebra.

   (a) $a(a + b) = a$.

   (b) $a(\bar{a} + b) = ab$.

10. Simplify each of the following Boolean algebra expressions as much as possible. You might find it useful to draw Venn diagrams first.

    (a) $xy + x\bar{y}$.

    (b) $x\bar{y}\bar{x} + xy\bar{x}$.

    (c) $xy + x\bar{y} + \bar{x}y$.

    (d) $xyz + xy\bar{z} + x\bar{y}z + x\bar{y}\bar{z}$.

    (e) $\bar{x}\bar{y}z + x\bar{y}\bar{z} + x\bar{y}z$.

    (f) $x + yz + \bar{x}y + \bar{y}xz$.

11. Let $B = \{1, 2, 3, 5, 6, 10, 15, 30\}$. Let $m, n \in B$. Define $m \cdot n$ to be the greatest common divisor of m and n; that is, the largest whole number that divides both m and n exactly. Define $m + n$ to be the lowest common multiple of m and n; that is, the smallest whole number into which both m and n divide exactly. Define $\mathbf{0} = 1$ and define $\mathbf{1} = 30$.

    (a) Show that with these definitions B is a Boolean algebra.

    (b) How is this Boolean algebra related to the Boolean algebra of all subsets of the set $\{2, 3, 5\}$?

    (c) How do you compute complements of elements?

12. Let $X = \{0, 1\}$ be *any* two-element Boolean algebra. Prove that $+$ must behave like $\vee$, that $\cdot$ must behave like $\wedge$ and that complementation must behave like negation.

13. Prove that
$$\overline{(\bar{x} + y)} + y = \overline{(\bar{y} + x)} + y$$

    in any Boolean algebra. *This result links Boolean algebras with more general algebraic structures called* MV *algebras which are the subject of contemporary research. The initials 'MV' stand for 'many-valued' and derive from* many-valued logic *where there are more than two truth values.*

14. A set S has a binary operation $+$, meaning that $+$ maps an ordered pair to the outcome $a + b$, and a unary operation $'$, meaning that each element a is mapped to the element a', such that the following properties are satisfied:

    (a) $a + (b + c) = (a + b) + c$.

    (b) $a + b = b + a$.

(c) $(a' + b')' + (a' + b)' = a$.

Define $a \cdot b = (a' + b')'$, $1 = a + a'$, for any $a \in S$, and $0 = 1'$. Prove that $(S, +, \cdot, ', 0, 1)$ is a Boolean algebra. *I don't think the word 'hard' does justice to this question. See [35, 36] if you get stuck.*

2.3 Combinational circuits

In this section and the next, we shall describe how Boolean algebra can be used to help design the kinds of circuits important in building computers. The discovery that Boolean algebras could be used in this way is due to Claude Shannon (1916–2001) in his MSc thesis of 1937.[4] The relationship between Boole's pioneering work in pure mathematics and Shannon's application of that work to electrical engineering is described in [58].

A computer circuit is a physical, not a mathematical, object. It is constructed out of transistors, resistors and capacitors and the like and powered by electricity. We shall not be dealing with such real circuits in this book. Instead, we shall describe some simple mathematical models which are idealizations of such circuits. Nevertheless, the models we describe are the basis of designing real circuits, and their theory is described in all books on digital electronics. Currently, computers work using binary logic, meaning that their circuits operate using two values of some physical property, such as voltage. This feature will be modelled by the two elements of the Boolean algebra $\mathbb{B}$. We shall work with mathematical expressions involving letters such as x, y, z; these are *Boolean variables* meaning that $x, y, z \in \mathbb{B}$. Circuits come in two types: *combinational circuits*, which have no internal memory, and *sequential circuits*, which do. In this section, we describe combinational circuits which are, to a first mathematical approximation, nothing other than *Boolean functions* $f \colon \mathbb{B}^m \to \mathbb{B}^n$. I should add that real combinational circuits have features that are not captured by our simple model; these are discussed in books on digital electronics. Sequential circuits are discussed in the next section.

2.3.1 How gates build circuits

Let C be a combinational circuit with m input wires and n output wires. We can think of C as consisting of n combinational circuits $C_1, \ldots, C_n$, each having m input wires but only one output wire each. The combinational circuit C_i tells us about how the ith output of the circuit C behaves with respect to inputs. These n combinational circuits can be combined into one circuit, possibly very inefficiently, by using the operation of 'fanout' described below.

---

[4]Shannon went on to lay the foundations of information theory; his work underpins the role of cryptography used in protecting sensitive information.

Thus it is enough to describe combinational circuits with m inputs and 1 output. Such a circuit is said to describe a *Boolean function* $f\colon \mathbb{B}^m \to \mathbb{B}$ where $\mathbb{B}^m$ represents all the possible 2^m inputs. Boolean functions with one output are described by means of an *input/output table*. Here is an example of such a table.

x	y	z	$f(x,y,z)$
1	1	1	0
1	1	0	0
1	0	1	0
1	0	0	1
0	1	1	0
0	1	0	1
0	0	1	1
0	0	0	0

Our goal is to show that any such Boolean function can be constructed from certain simpler Boolean functions called *gates*. There are a number of different kinds of gates but we begin with the three basic ones.

The *and-gate* is the function $\mathbb{B}^2 \to \mathbb{B}$ defined by $(x, y) \mapsto x \cdot y$. We use the following symbol to represent this function:

The *or-gate* is the function $\mathbb{B}^2 \to \mathbb{B}$ defined by $(x, y) \mapsto x + y$. We use the following symbol to represent this function:

Finally, the *not-gate* is the function $\mathbb{B} \to \mathbb{B}$ defined by $x \mapsto \bar{x}$. We use the following symbol to represent this function:

Diagrams constructed using gates are called *circuits* and show how Boolean functions can be computed as we shall see. Such mathematical circuits can be converted into physical circuits with gates being constructed from simpler circuit elements called transistors which operate like electronic switches.

Example 2.3.1. Here is a simple circuit.

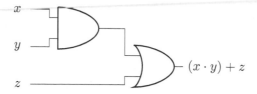

$$(x \cdot y) + z$$

Example 2.3.2. Because of the associativity of $\cdot$, the circuit

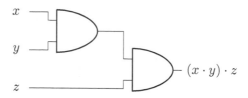

$$(x \cdot y) \cdot z$$

and the circuit

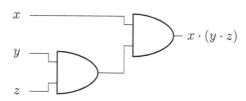

$$x \cdot (y \cdot z)$$

compute the same function. Similar comments apply to the operation $+$.

The main theorem in circuit design is the following. It is nothing other than the Boolean algebra version of Theorem 1.7.3, the result that says that every truth function arises as the truth table of a wff.

Theorem 2.3.3 (Fundamental theorem of circuit design). *Every Boolean function* $f \colon \mathbb{B}^m \to \mathbb{B}$ *can be constructed from and-gates, or-gates and not-gates.*

Proof. Assume that f is described by means of an input/output table. We deal first with the case where f is the constant function to 0. In this case,

$$f(x_1, \ldots, x_m) = (x_1 \cdot \overline{x_1}) \cdot x_2 \cdot \ldots \cdot x_m.$$

Next we deal with the case where the function f takes the value 1 exactly once. Let $\mathbf{a} = (a_1, \ldots, a_m) \in \mathbb{B}^m$ be such that $f(a_1, \ldots, a_m) = 1$. Define $\mathbf{m} = y_1 \cdot \ldots \cdot y_m$, called the *minterm* associated with $\mathbf{a}$, as follows:

$$y_i = \begin{cases} x_i & \text{if } a_i = 1 \\ \overline{x_i} & \text{if } a_i = 0. \end{cases}$$

Then $f(\mathbf{x}) = y_1 \cdot \ldots \cdot y_m$.

Finally, we deal with the case where the function f is none of the above. Let the inputs where f takes the value 1 be $\mathbf{a}_1, \ldots, \mathbf{a}_r$, respectively. Construct the corresponding minterms $\mathbf{m}_1, \ldots, \mathbf{m}_r$, respectively. Then

$$f(\mathbf{x}) = \mathbf{m}_1 + \ldots + \mathbf{m}_r. \qquad \square$$

Example 2.3.4. We illustrate the proof of Theorem 2.3.3 by means of the following input/output table.

x	y	z	$f(x, y, z)$
1	1	1	0
1	1	0	0
1	0	1	0
1	0	0	1
0	1	1	0
0	1	0	1
0	0	1	1
0	0	0	0

The three elements of $\mathbb{B}^3$ where f takes the value 1 are $(1, 0, 0)$, $(0, 1, 0)$ and $(0, 0, 1)$. The minterms corresponding to each of these inputs are $x \cdot \bar{y} \cdot \bar{z}$, $\bar{x} \cdot y \cdot \bar{z}$ and $\bar{x} \cdot \bar{y} \cdot z$, respectively. It follows that

$$f(x, y, z) = x \cdot \bar{y} \cdot \bar{z} + \bar{x} \cdot y \cdot \bar{z} + \bar{x} \cdot \bar{y} \cdot z.$$

We could, if we wished, attempt to simplify this Boolean expression. This becomes important when we wish to convert it into a circuit.

Example 2.3.5. The input/output table below

x	y	$x \oplus y$
1	1	0
1	0	1
0	1	1
0	0	0

defines *exclusive or* (or *xor*). By Theorem 2.3.3, we have that

$$x \oplus y = \bar{x} \cdot y + x \cdot \bar{y}.$$

We may describe this algebraic expression by means of a parse tree just as we did in the case of wff in Section 1.2:

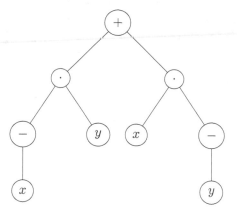

This parse tree may be converted into a circuit in a series of stages but requires two further circuit elements as follows:

- First, we require two input wires: one labelled x and one labelled y. But in the parse tree x occurs twice and y occurs twice. We therefore need a new circuit element called *fanout*. This has one input and then branches, with each branch carrying a copy of the input. In this case, we need one fanout with input x and two outward edges and another with input y and two outward edges.

- In addition, we need to allow wires to cross but not to otherwise interact. This is called *interchange* and forms the second additional circuit element we need.

Finally, we replace the Boolean symbols by the corresponding gates and rotate the diagram ninety degrees clockwise so that the inputs come in from the left and the output emerges from the right. We therefore obtain the following circuit diagram:

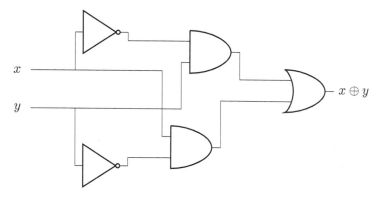

For our subsequent examples, it is convenient to abbreviate the circuit for xor by means of a single circuit symbol called an *xor-gate*.

$$x \oplus y$$

Example 2.3.6. There are two further gates that you are likely to encounter. The first is the *nor-gate*:

$$x \text{ nor } y$$

This has the following input/output table

x	y	x nor y
1	1	0
1	0	0
0	1	0
0	0	1

and is just the Boolean algebra version of the PL connective **nor**. The second is the *nand-gate*:

$$x \text{ nand } y$$

This has the following input/output table

x	y	x nand y
1	1	0
1	0	1
0	1	1
0	0	1

and is just the Boolean algebra version of the PL connective **nand**.

The following theorem is nothing other than the Boolean algebra version of Proposition 1.6.4.

Theorem 2.3.7 (Nor-gates and nand-gates).

1. *Every Boolean function $f \colon \mathbb{B}^m \to \mathbb{B}$ can be constructed from only nor-gates.*

2. *Every Boolean function $f \colon \mathbb{B}^m \to \mathbb{B}$ can be constructed from only nand-gates.*

The hardware of a computer is the part that we tend to notice first. It is also the part that most obviously reflects advances in technology. The single most important component of the hardware of a computer is called a *transistor*. These days, computers contain billions of transistors. There is even an empirical result, known as *Moore's law*, which says the number of transistors in integrated circuits doubles roughly every two years. Transistors were invented in the 1940's and are constructed from semiconducting materials, such as silicon. The way they work is complex and depends on the quantum-mechanical properties of semiconductors, but what they *do* is very simple. The basic function of a transistor is to amplify a signal. A transistor is constructed in such a way that a weak input signal is used to control a large output signal and so achieve amplification of the weak input. An extreme case of this behaviour is where the input is used to turn the large input on or off. That is, where the transistor is used as an electronic switch. It is this function which is the most important use of transistors in computers. From a mathematical point of view, a transistor can be regarded as a device with one input, one control (input) and one output:

- If 0 is input to the control then the switch is closed and the output is equal to the input.

- If 1 is input to the control then the switch is open and the output is 0 irrespective of the input.

A transistor on its own doesn't appear to accomplish very much, but more complicated behaviour can be achieved by connecting transistors together into circuits. Because of De Morgan's laws for Boolean algebras, Proposition 2.2.10, every Boolean expression is equal to one in which only $\cdot$ and $^-$ appear. We shall prove that and-gates and not-gates can be constructed from transistors, and so we will have proved that every combinational circuit can be constructed from transistors.

I shall regard the transistor as a new Boolean operation that I shall write as $x \,\square\, y$. This is certainly not standard notation (there is none), but we only need it in this section. Its input/output behaviour is described by the following table.

x	y	$x \,\square\, y$
1	1	0
1	0	1
0	1	0
0	0	0

Observe that $x \,\square\, y \neq y \,\square\, x$. Clearly, $x \,\square\, y = x \cdot \bar{y}$. We now carry out a couple of calculations.

1. $1 \,\square\, y = 1\bar{y} = \bar{y}$. Thus by fixing $x = 1$ we can negate y.

2. Observe that $x \cdot y = x \cdot \bar{\bar{y}}$. Thus

$$x \cdot y = x \,\square\, \bar{y} = x \,\square\, (1 \,\square\, y).$$

We have therefore proved the following.

Theorem 2.3.8 (The fundamental theorem of transistors). *Every combinational circuit can be constructed from transistors.*

2.3.2   A simple calculator

The goal of this section is to show that we can actually build something useful with the theory we have developed so far. We shall construct, in stages, a circuit that will add two four-bit numbers together. In everyday life, we write numbers down using base ten but in computers, it is more natural to treat numbers as being in base two or *binary*. We begin by explaining what this means. Recall that a string is simply an ordered sequence of symbols. If the symbols used in the string are taken only from $\{0, 1\}$, the set of *binary digits*, then we have a *binary string*. All kinds of data can be represented by binary strings but here we are only interested in the way they can be used to encode natural numbers. The key idea is that every natural number can be represented as a sum of powers of two. The following example illustrates how to do this. Recall that $2^0 = 1$, $2^1 = 2$, $2^2 = 4$, $2^3 = 8$, ...

Example 2.3.9. Write 316 as a sum of powers of two.

- We first find the highest power of 2 that is less than or equal to our number. We see that $2^8 < 316$ but $2^9 > 316$. We can therefore write $316 = 2^8 + 60$.

- We now repeat this procedure with 60. We find that $2^5 < 60$ but $2^6 > 60$. We can therefore write $60 = 2^5 + 28$.

- We now repeat this procedure with 28. We find that $2^4 < 28$ but $2^5 > 28$. We can therefore write $28 = 2^4 + 12$.

- We now repeat this procedure with 12. We find that $2^3 < 12$ but $2^4 > 12$. We can therefore write $12 = 2^3 + 4$. Of course $4 = 2^2$.

It follows that
$$316 = 2^8 + 2^5 + 2^4 + 2^3 + 2^2,$$
a sum of powers of two.

Once we have written a number as a sum of powers of two we can encode that information as a binary string. How to do so is illustrated by the following example that continues the one above.

Example 2.3.10. We have that
$$316 = 2^8 + 2^5 + 2^4 + 2^3 + 2^2.$$

We now set up the following table

2^8	2^7	2^6	2^5	2^4	2^3	2^2	2^1	2^0
1	0	0	1	1	1	1	0	0

which includes all powers of two up to and including the largest one used. We say that the binary string

$$100111100$$

is the *binary representation* of the number 316.

Given a number written in binary it is a simple matter to convert it back into base ten.

Example 2.3.11. We show how to convert the number 110010 written in base two into a number written in base ten. The first step is to draw up a table of powers of two.

2^5	2^4	2^3	2^2	2^1	2^0
1	1	0	0	1	0

We now add together the powers of two which correspond to a '1'. This is just $2^5 + 2^4 + 2^1 = 50$ which is the corresponding number in decimal.

We have shown how every natural number can be written in base two. In fact, all of arithmetic can be carried out working solely in base two although we shall only need to describe how to do addition.

> *At this point, there is some notational congestion since $+$ is the Boolean operation we have called addition which is different from addition in the arithmetical sense. I shall therefore use the word 'plus' to mean arithmetic addition.*

The starting point is to consider how to add two one-bit binary numbers together. The table below shows how this is done:

x	y	carry(x, y)	sum(x, y)
1	1	1	0
1	0	0	1
0	1	0	1
0	0	0	0

Thus 1 plus $1 = 10$. However, this is not quite enough to enable us to add two arbitrary binary numbers together. The algorithm for doing this is the same as in base ten except that two is carried rather than ten. Because of the carries, it is necessary to know how to add three binary digits rather than just two. The table below shows how:

x	y	z	carry(x, y, z)	sum(x, y, z)
1	1	1	1	1
1	1	0	1	0
1	0	1	1	0
1	0	0	0	1
0	1	1	1	0
0	1	0	0	1
0	0	1	0	1
0	0	0	0	0

The above table can be used to add two binary numbers together as the following example illustrates.

Example 2.3.12. We calculate 11 plus 1101 in binary using the second of the two tables above. We first pad out the first string with 0's to get 0011. We now write the two binary numbers one above the other; their sum will appear in the third row; the fourth row in our calculations is for the carry bits but our rightmost calculation has no carry bit and so we record this as a 0.

0	0	1	1
1	1	0	1
			0

Now work from right-to-left a column at a time adding up the digits you see and making any necessary carries. Observe that in the first column on the right there is no carry but it is more helpful, as we shall see, to think of this as a 0 carry. Here is the sequence of calculations:

0	0	1	1
1	1	0	1
			0
		1	0

0	0	1	1
1	1	0	1
		0	0
	1	1	0

0	0	1	1
1	1	0	1
	0	0	0
1	1	1	0

0	0	1	1
1	1	0	1
1	0	0	0	0
1	1	1	1	0

We find that the sum of these two numbers is 10000.

We shall describe in stages how to build a circuit that will add up two four-bit numbers in binary.

Example 2.3.13. Our first circuit is known as a *half-adder* which has two inputs and two outputs and is defined by the following input/output table:

x	y	c	s
1	1	1	0
1	0	0	1
0	1	0	1
0	0	0	0

where we have written c instead of $\text{carry}(x, y)$ and s instead of $\text{sum}(x, y)$. This treats the input x and y as numbers in binary and then outputs their sum. Observe that $s = x \oplus y$ and $c = x \cdot y$. Thus we may easily construct a circuit that implements this function:

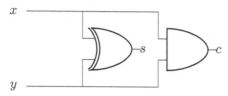

In what follows, it will be useful to have a single symbol representing a half-adder. We use the following:

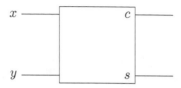

But as we have seen, the half-adder will not be quite enough for what we need.

Example 2.3.14. Our next circuit is known as a *full-adder* which has three inputs and two outputs and is defined by the following input/output table:

x	y	z	c	s
1	1	1	1	1
1	1	0	1	0
1	0	1	1	0
1	0	0	0	1
0	1	1	1	0
0	1	0	0	1
0	0	1	0	1
0	0	0	0	0

where we have written c instead of $\text{carry}(x, y, z)$ and s instead of $\text{sum}(x, y, z)$. This treats the three inputs as numbers in binary and adds them together. The following circuit realizes this behaviour using two half-adders completed with an or-gate.

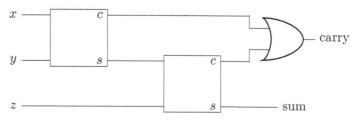

To understand why this circuit works, we make the following observations which can all be verified either by algebra or by checking against the table for the full-adder:

- $\mathrm{sum}(x, y, z) = \mathrm{sum}(\mathrm{sum}(x, y), z)$.

- $\mathrm{carry}(x, y, z) = \mathrm{carry}(x, y) + \mathrm{carry}(\mathrm{sum}(x, y), z)$, where $+$ is the Boolean operation not addition of binary numbers.

- $\mathrm{carry}(x, y)$ and $\mathrm{carry}(\mathrm{sum}(x, y), z)$ are never both equal to 1 for the same values of x and y.

We shall represent the full-adder by means of the following box-diagram:

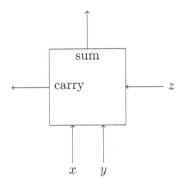

We now have all the ingredients we need to build a circuit that will add together two four-bit binary numbers.

Example 2.3.15. Full-adders are the building blocks from which all arithmetic computer circuits can be built. Specifically, suppose that we want to add two four-bit binary numbers together where we pad the numbers out by adding 0 at the front if necessary. Denote them by $m = a_3a_2a_1a_0$ and $n = b_3b_2b_1b_0$. The sum $m + n$ in base two, which will be denoted by m plus $n = c_4c_3c_2c_1c_0$, is computed in a similar way to calculating a sum in base ten. Thus first calculate a_0 plus b_0, write down the sum bit, c_0, and pass any carry to be added to a_1 plus b_1 and so on. Although a_0 plus b_0 can be computed by a half-adder subsequent additions may require the addition of three bits because of the presence of a carry bit. For this reason, we actually use four full-adders joined in series with the rightmost full-adder having one of its inputs set to 0.

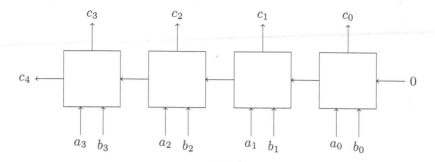

One point to observe in all of this is that we did not use Theorem 2.3.3 directly. Although this theorem guarantees that we *can* build any combinational circuit we want, to do so using the method employed in its proof would be prohibitively complicated except in small cases. Instead, we designed the circuit in stages taking what we constructed at one stage as a building block to be used in the next.

Exercises 2.3

---

1. Write down Boolean expressions for each of the following three input/output behaviours.

x	y	z	$f(x,y,z)$	$g(x,y,z)$	$h(x,y,z)$
1	1	1	0	1	0
1	1	0	0	0	1
1	0	1	0	0	1
1	0	0	0	0	0
0	1	1	0	0	1
0	1	0	1	0	0
0	0	1	0	0	0
0	0	0	0	1	0

2. Simplify each of the following combinational circuits as much as possible.

(a)

(b)

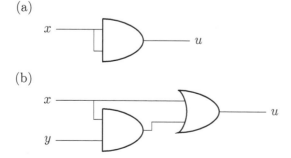

(c)

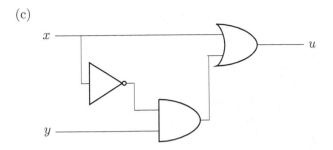

3. The following diagram shows a circuit with two inputs and one output. Write the output u as a Boolean expression in terms of the inputs x and y.

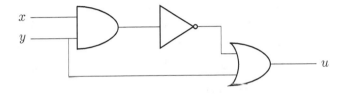

4. The following diagram shows a circuit with three inputs and one output. Write the output u as a Boolean expression in terms of the inputs x, y and z.

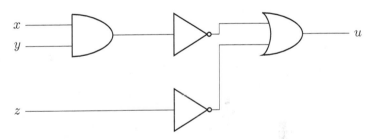

5. Convert the following numbers into binary.

   (a) 10.

   (b) 42.

   (c) 153.

   (d) 2001.

6. Convert the following binary numbers into decimal.

   (a) 111.

   (b) 1010101.

   (c) 111000111.

7. Carry out the following arithmetical additions in binary.

    (a) $11 + 11$.

    (b) $10110011 + 1100111$.

    (c) $11111 + 11011$.

8. Show how to construct a nor-gate from transistors.

9. A *Fredkin gate* has three inputs, a, b, c and three outputs, a', b', c'. It behaves as follows where $c' = c$. If $c = 0$ then $a' = a$ and $b' = b$. If $c = 1$ then $a' = b$ and $b' = a$.

    (a) Draw up an input/output table for this gate.

    (b) What happens if you connect two such gates in series?

    (c) Write down a Boolean expression for a' in terms of b and c if $a = 0$.

    (d) Write down a Boolean expression for a' in terms of c if $a = 1$ and $b = 0$.

    (e) Write down a Boolean expression for a' in terms of a and c if $b = 1$.

This question, and the next, come from the theory of reversible comput-ing *[59]*.

10. A *Toffoli gate* has three inputs a, b, c and three outputs a', b', c'. It be-haves as follows: $a' = a$, $b' = b$ and $c' = a \cdot b \oplus c$.

    (a) Draw up an input/output table for this gate.

    (b) What happens if you connect two such gates in series?

    (c) Write down the Boolean expression for c' in terms of a and b when $c = 0$.

    (d) Write down the Boolean expression for c' in terms of a when $b = 1$ and $c = 1$.

    (e) Write down the Boolean expressions for b' and c' in terms of b when $a = 1$ and $c = 0$. What is this circuit element?

2.4 Sequential circuits

Suppose that we want to add three four-bit binary numbers together. We can use the combinational circuit of the previous section to calculate the sum of two such binary numbers, but then we would have to remember that result and add it to the third four-bit binary number to calculate the total sum of the three numbers. It follows that to build a real calculator and, more expansively, a real computer, we shall need memory. But now time will play a role since if there is no time there can be no memory.

Example 2.4.1. I have drawn a picture below of a 'black box circuit'. It has one input wire and one output wire and your job is to figure out what it is doing. You are allowed to input bits and observe what comes out.

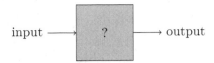

You begin by inputting a 0 and a 0 is output. You then input a 1 and a 0 comes out again. If this were a combinational circuit, you would then be able to say that the output was always 0. But, just to be sure, you input 0 again but to your surprise a 1 is output. Thus, whatever this circuit is, it is not a combinational one. Rather than continuing with further examples, I shall tell you exactly what this circuit is doing: it always outputs the previous input except for the first output which I initialized to 0 (I could equally well have initialized it to 0). It is our first example of a sequential circuit since it has a very limited form of memory: an extreme short-term memory, if you like. This particular circuit is called a *delay* and we shall use it as a building block in making more complex sequential circuits.

The concept of *state* plays an important role in understanding circuits with memory where a state is simply a memory configuration. Here is an example.

Example 2.4.2. The diagram below shows an array.

x_1	x_2
x_3	x_4

Each cell of the array can hold one bit. There are therefore $2^4 = 16$ such arrays

from
0	0
0	0
to
1	1
1	1
.

Each of these 16 possible arrays can be regarded as a particular memory or, as we shall say, a state. We encountered a concrete example of this situation where we used two coins to record 'memories'. In that case, there were four states: HH, HT, TH, TT.

A sequential circuit is a combinational circuit which has access to a memory. The following diagram is a schematic representation of how such circuits are to be regarded:

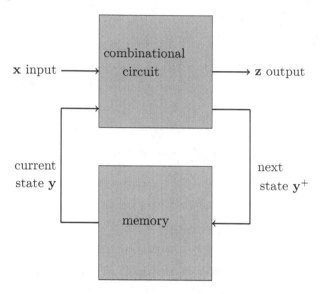

This diagram has one additional difference from a combinational circuit and that is *feedback*. This is represented by the loop that comes out of the combinational circuit, passes through the memory, and then goes back into the combinational circuit. Thus the current state and the input determine the next state and the output. *In what follows, we shall not include an explicit output but it is easy to include one.*

Example 2.4.3. We begin with an example to illustrate how the above schematic diagram is realized in practice. We start with the purely combinational circuit below:

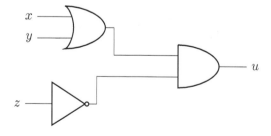

It is easy to check that $u = \bar{z} \cdot (x + y)$.

Alternatively, we can construct an input/output table to describe the behaviour of this combinational circuit:

x	y	z	u
1	1	1	0
1	1	0	1
1	0	1	0
1	0	0	1
0	1	1	0
0	1	0	1
0	0	1	0
0	0	0	0

I shall now relabel the above circuit and interpret its behaviour in a different way.

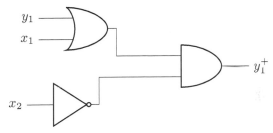

The ordered pair (x_1, x_2) of Boolean variables x_1 and x_2 is now interpreted as the input whereas the Boolean variable y_1 is interpreted as the *current state* and y_1^+ is interpreted as the *next state*. We can still use an input/output table to describe the behaviour of this circuit:

y_1	x_1	x_2	y_1^+
1	1	1	0
1	1	0	1
1	0	1	0
1	0	0	1
0	1	1	0
0	1	0	1
0	0	1	0
0	0	0	0

But now we can also use a *state transition diagram*. This is constructed directly from the above table. First, we describe the input alphabet. This consists of all the elements of $\mathbb{B}^2$ which we label as follows for convenience:

$$a = (0,0), \quad b = (1,0), \quad c = (0,1), \quad d = (1,1).$$

Thus our input alphabet is $\{a, b, c, d\}$. There are two states corresponding to the two possible values of y_1. We now use the above table to construct the following finite-state automaton; these were introduced in Section 1.5.2:

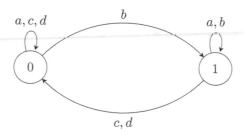

It is important to observe that the input/output table can be reconstituted from this diagram. If we remove inputs a, d then our finite-state automaton becomes more symmetric:

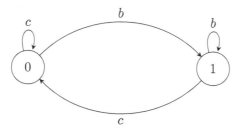

Observe that when c is applied, the automaton always goes into state 0 whereas when b is applied it always goes into state 1. Thus this automaton is capable of remembering one bit. Because of the way it functions, it is called a *flip-flop*.

I have used y_1 and y_1^+ as *memory variables* where y_1^+ can be read as *next* y_1. However, we still don't quite have a real circuit. Given what we are doing, we could redraw it by connecting y_1^+ to y_1 but the drawback of this is that you have to know how to handle the feedback correctly. To make this clear, I introduce an explicit *delay gate* whose function is to remember for one unit of time:

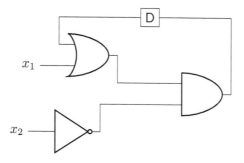

This serves to demarcate between the current state on the left-hand side and the next state on the right-hand side.

What follows is tangential to the main text which is why I have placed it in a smaller font. We analyse the above example from a mathematical perspective. Our starting point was a Boolean function $F: \mathbb{B}^3 \to \mathbb{B}$ which can equally well be described by means of a combinational circuit. In this case, $u = F(x, y, z) = \bar{z} \cdot (x + y)$. Essentially, such functions are timeless. However, in our sequential machine, y and z are replaced by x_1 and x_2, respectively, where now $x_1: \mathbb{N} \to \mathbb{B}$ and $x_2: \mathbb{N} \to \mathbb{B}$ are really functions where the domains of these functions are the natural numbers which represent the values of discrete time; in concrete terms, $x_1(t)$ is the value of the x_1-input at time t, and similarly for $x_2(t)$. Both x_1 and x_2 are known functions. We now define a new function $y_1: \mathbb{N} \to \mathbb{B}$ by means of the following equation

$$y_1(t+1) = \overline{x_2(t)} \cdot (y_1(t) + x_1(t)).$$

At first, this looks problematic because y_1 appears on both sides. But there is a difference. On the right-hand side, we evaluate y_1 at time t whereas on the left-hand side we evaluate y_1 at time $t + 1$. If we know the value of y_1 at $t = 0$ then we can calculate the value of y_1 at any subsequent time using the above equation. This is an example of defining a function by *recursion*. In our case, this recursion simplifies because y_1, x_1, x_2 can each only assume a finite number of values. This means that we can, in fact, forget about the time variable t and work instead with the operator $y_1 \mapsto y_1^+$. Thus our equation above can be written simply

$$y_1^+ = \overline{x_2} \cdot (y_1 + x_1)$$

which in turn can be encoded by means of a finite-state automaton.

Let me now describe in general the relationship between Boolean recursions and sequential circuits. A set of *Boolean recursions* is a set of r Boolean equations of the following form:

$$y_1^+ \quad = \quad B_1(x_1, \ldots, x_s, y_1, \ldots, y_r)$$
$$\cdots$$
$$y_r^+ \quad = \quad B_r(x_1, \ldots, x_s, y_1, \ldots, y_r).$$

The right-hand sides of each of these equations will be the Boolean $+$ of minterms constructed from the $r + s$ Boolean variables $x_1, \ldots, x_s, y_1, \ldots, y_r$. In our example above, we had just one such Boolean recursion.

To construct the sequential circuit associated with these Boolean recursions, proceed as follows. First, construct a combinational circuit from the Boolean recursions. Define the input variables to be the ordered pair consisting of $\mathbf{x}$, the ordered s-tuples $(x_1, \ldots, x_s)$, and $\mathbf{y}$, the ordered r-tuples $(y_1, \ldots, y_r)$; define the output variables to be $\mathbf{y}^+$, the ordered r-tuples $(y_1^+, \ldots, y_r^+)$. Second, construct the sequential circuit obtained from the above combinational circuit by connecting the output labelled y_i^+ to the input labelled y_i by means of a delay.

It should now be clear that sequential circuits and Boolean recursions are two ways of describing the same thing, though Boolean recursions are typographically easier to write down.

Now we turn to the relationship between Boolean recursions and finite state automata. Let me begin by defining precisely what is meant by a finite state automaton $\mathbf{A} = (Q, A, \delta)$. It consists of a finite set $Q = \{q_1, \ldots, q_m\}$ of states, a finite input alphabet $A = \{a_1, \ldots, a_n\}$ and a state transition function $\delta \colon Q \times A \to Q$. Finite state automata are usually represented as directed graphs whose directed edges are labelled by elements of the input alphabet: the vertices are the states and there is a directed edge labelled with a from the vertex labelled q to the vertex labelled q' if $\delta(q, a) = q'$.

We now show how we can construct a sequential circuit/Boolean recursions that describes the finite state automaton. First, we have to code the m states by means of elements of $\mathbb{B}^r$, though we shall use binary *strings* of length r. If $m = 2^r$ then the states can be labelled by all the binary strings of length r from $\overbrace{0 \ldots 0}^{r}$ to $\overbrace{1 \ldots 1}^{r}$ whereas if $m < 2^r$ and 2^r is the smallest power of two greater than or equal to m then we simply omit the unneeded binary strings. It is clear that there are many ways in which this labelling can be carried out. From our point of view, it doesn't matter how you do this. Next, we have to code the n elements of the input alphabet. This is done in the same way as the coding of the states, so if $n = 2^s$ then we encode the elements of the input alphabet by all the binary strings of length s from $\overbrace{0 \ldots 0}^{s}$ to $\overbrace{1 \ldots 1}^{s}$ whereas if $n < 2^s$ and 2^s is the smallest power of two greater than or equal to n then we simply omit the unneeded binary strings. Once again, there will be considerable choice in how this is done. The upshot is that the sequential circuit/Boolean recursions we construct from the finite state automaton will not be unique. Denote by $\mathbf{c}(q)$, the binary string associated with the state q, and $\mathbf{c}(a)$, the binary code associated with the input letter a. We now construct an input/output table where each row has the following form:

$$\boxed{\mathbf{c}(q) \mid \mathbf{c}(a) \parallel \mathbf{c}(\delta(q, a))}$$

for all $q \in Q$ and $a \in A$. To understand how to read this table and so use it to construct a sequential circuit, focus first on the column labelled $\mathbf{c}(a)$. This is s bits long and so the sequential circuit will have s input wires. If we write $\mathbf{c}(a) = x_1 \ldots x_s$ then these wires are labelled $x_1, \ldots, x_s$. Let $\mathbf{c}(q) = y_1 \ldots y_r$. Then our input/output table has the following $2r + s$ columns

$$\boxed{y_1 \ldots y_r \mid x_1 \ldots x_s \parallel y_1^+ \ldots y_r^+}$$

This can be regarded as the input/output table of a combinational circuit with $r + s$ inputs, $y_1, \ldots, y_r, x_1, \ldots, x_s$, and r outputs, $y_1^+, \ldots, y_r^+$. This circuit can be constructed using the methods of this chapter. Finally, each output labelled y_i^+ should be connected to the input labelled y_i by means of a delay to obtain the sequential circuit representing the original finite-state automaton.

Example 2.4.4. We illustrate the above process by means of an example discussed in [51]. We begin with a finite-state automaton; I have already carried out an encoding of the states by means of binary strings.

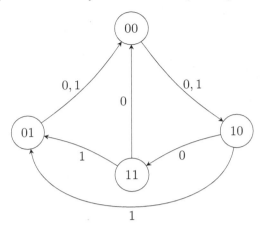

We now construct an input/output table. The input variable is x and the states are binary strings $y_1 y_2$.

y_1	y_2	x	y_1^+	y_2^+
0	0	0	1	0
0	0	1	1	0
0	1	0	0	0
0	1	1	0	0
1	0	0	1	1
1	0	1	0	1
1	1	0	0	0
1	1	1	0	1

We treat this as a combinational circuit with inputs y_1, y_2, x and outputs y_1^+, y_2^+. You can check that the following two equations describe this behaviour exactly:

$$y_1^+ = \bar{y}_1\, \bar{y}_2 + \bar{x}\, \bar{y}_2$$
$$y_2^+ = y_1 \bar{y}_2 + x y_1.$$

It is now routine to construct the corresponding combinational circuit and, then, the corresponding sequential circuit.

Exercises 2.4

1. For each of the following Boolean recursions, construct a finite state automaton with two states, y, and input, x.

   (a) $y^+ = x + y$.

   (b) $y^+ = \bar{x} + y$.

   (c) $y^+ = x + \bar{y}$.

   (d) $y^+ = \bar{x} + \bar{y}$.

   (e) $y^+ = x \cdot y$.

   (f) $y^+ = x \cdot \bar{y} + \bar{x} \cdot y$.

2. In this question, denote the input by x and the state variable by y. Construct (a) an input/output table for the following finite-state automaton and then (b) a Boolean recursion expressing y^+ in terms of x and y.

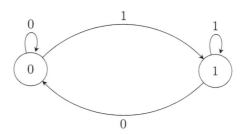

3. In this question, denote the input by x and the state variables by y_1 and y_2. Construct (a) an input/output table for the following finite-state automaton and then (b) the Boolean recursions expressing y_1^+ and y_2^+ in terms of x, y_1 and y_2.

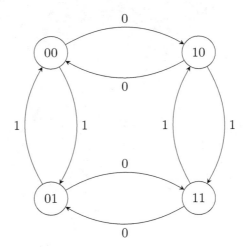

Chapter 3

First-order logic

> *What we cannot speak about, we must pass over in silence.* —
> Ludwig Wittgenstein.

The following is a famous[1] argument: (1) All men are mortal. (2) Socrates is a man. (3) ∴ Socrates is mortal. Here (1) and (2) are the assumptions and (3) is the conclusion. If you agree to the truth of (1) and (2) then you are obliged to accept the truth of (3). This is therefore a valid argument, and not a very complex one, but it cannot be shown to be valid using propositional logic. Thus propositional logic, whilst very useful, is not up to the job of analysing even quite simple arguments. The goal of this last chapter is to add features to propositional logic that will make it more powerful and more useful; in particular, we shall see that in this richer logic we can prove that the above argument is valid. There will be a price to pay in that the resulting system will be intrinsically harder to work with.

We shall study what is known as *first-order logic.*[2] Specifically,

> first-order logic = PL + predicates + functions + quantifiers.

In this chapter, however, to keep the presentation manageable, our formal logic will not use the full generality offered by functions though we shall allow constants. Logic has been described as the 'calculus of computer science', meaning that it is as indispensible to computer science as calculus is to physics. The role of logic in mathematics is even more fundamental being described by the following equation:

> Mathematics = Set Theory + FOL.

---

[1] Famous it might be, but locatable it isn't. I assumed it would be in Aristotle but this is apparently wrong. It seems to occur first in the work of William of Ockham. See https://philosophy.stackexchange.com/questions/34461/all-men-are-mortal-socrates-is-a-man-therefore-socrates-is-mortal-original?utm_medium=organic&utm_source=google_rich_qa&utm_campaign=google_rich_qa. My thanks to Jamie Gabbay for raising this question.

[2] The term *first-order* has a technical meaning: it means that quantifiers can only range over variables. There is such a thing as *second-order* logic where quantification is allowed over variables that designate relations or functions as well, but that belongs in a second course on logic, rather than a first. And yes, there are also *higher-order* logics.

3.1 First steps

First-order logic is a more complex system than propositional logic so to ease learning it we shall proceed in a way that combines the informal with the formal rather in the way we learn languages at school.

3.1.1 Names and predicates

Recall that a statement is a sentence which is capable of being either true or false. In PL, statements can only be analysed in terms of the usual PL connectives until we get to the atoms but the atoms themselves cannot then be analysed further. We shall now show that atoms can be split by means of two new ideas: names and predicates.

Names

A *name* picks out a specific individual. For example, 2, π, Darth Vader and Socrates are all names. Once something has been named, we can refer to it by using that name. Names seem simple but they play a leading role in FOL.[3]

Example 3.1.1. A nice example of the power of names occurs when students meet complex numbers for the first time. There is no *real number* whose square is -1 so we introduce the name i for the *entity* — which leaves open the question of just what it might be — defined to have the property that $i^2 = -1$ (together with a few other simple properties). Suspending belief about whether such an entity really exists, it becomes possible to prove some very interesting results in mathematics. Of course, you cannot just call things into existence by naming them — otherwise both unicorns and the Loch Ness Monster would exist — the key feature of mathematical existence is that *no contradictions arise*; this is the idea that lies behind truth trees for first-order logic.

> In mathematics, we tend to use the word 'constant' rather than the word 'name'. I will use both words to mean the same thing depending on the context: informally, I shall use the word 'name' whereas formally I shall use the word 'constant'.

Predicates

We begin by again analysing natural languages but this time we focus on the internal structure of sentences. At its simplest, a sentence can be analysed into a *subject* and a *predicate*. For example, the sentence 'grass is green' has

---

[3]There is a tarn near the summit of Haystacks in the Lake District that had no name until Alfred Wainwright dubbed it *Innominate Tarn*.

'grass' as its subject and 'is green' as its predicate. To make the nature of the predicate clearer, we might write '— is green' to indicate the slot into which a name could be fitted. Or, more formally, we could use a *variable* and write 'x is green'. In addition, we could symbolize this predicate by writing $G(x) = $ 'x is green'. If we replace the variable x by a name we get a statement that may or may not be true. Thus $G(\text{grass})$ is the statement 'grass is green', which is true, whereas $G(\text{the sky})$ is the statement 'the sky is green', which is false. This is an example of a *1-place predicate* because there is exactly one slot into which a name can be placed to yield a statement. Other examples of 1-place predicates are: 'x is a prime', 'x likes honey', 'x is mortal' and 'x is a man'.[4] There are also *2-place predicates*. For example,

$$P(x, y) = \text{'}x \text{ is the father of } y\text{'}$$

is such a predicate because there are two slots that can be replaced by names to yield statements. Thus, *spoiler alert*, $P(\text{Darth Vader}, \text{Luke Skywalker})$ is true but $P(\text{Winnie-the-Pooh}, \text{Darth Vader})$ is false. More generally, for any finite n,

$$P(x_1, \ldots, x_n)$$

denotes an *n-place predicate* or an *n-ary* predicate.[5] Here $x_1, \ldots, x_n$ are n variables that mark the n positions into which names can be slotted. There is no limit to the arities of the predicates that you might meet. For example, $F(x_1, x_2, x_3)$ is the 3-place predicate 'x_1 fights x_2 with an x_3'. By inserting names, we get the statement

$$F(\text{Luke Skywalker}, \text{Darth Vader}, \text{lightsaber}).$$

To keep things simple, I shall mostly use 1-place predicates and 2-place predicates in examples.

Atomic formulae

In FOL, names are called *constants* and are usually denoted by the rather more mundane $a, b, c, \ldots$ or $a_1, a_2, a_3 \ldots$ though for 'technical reasons' we might want some of the constants to be chosen from a set U, say. *Variables* are denoted by $x, y, z \ldots$ or $x_1, x_2, x_3, \ldots$; they have no fixed meaning but serve as place-holders into which constants can be slotted. An *atomic formula* is a predicate whose slots are filled with either variables or constants. We may adapt parse trees to the more general logic we are developing. The atomic formula $P(x_1, x_2, a)$, for example, gives rise to the following parse tree:

---

[4]A logic, in which the only predicates that appear are 1-place predicates, is said to be *monadic*.

[5]We also say that the predicate has an *arity* of n.

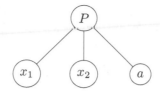

3.1.2 Relations

We now turn to the question of what predicates do. 1-place predicates describe sets.[6] Thus the 1-place predicate $P(x)$ describes the set

$$P = \{a \colon P(a) \text{ is true}\}$$

although this is usually written simply as

$$P = \{a \colon P(a)\}.$$

For example, if $P(x)$ is the 1-place predicate 'x is a prime' then the set it describes is the set of prime numbers. To deal with 2-place predicates, we use ordered pairs. Let D be any non-empty set. Denote by D^2 the set of all ordered pairs (a, b) where $a, b \in D$. Now let $P(x, y)$ be a 2-place predicate defined on D. Then it describes the set

$$P = \{(a, b) \colon P(a, b) \text{ is true}\} \subseteq D^2$$

which is usually just written

$$P = \{(a, b) \colon P(a, b)\}.$$

A set of ordered pairs whose elements are taken from some set D is called a *binary relation* on the set D. Thus 2-place predicates describe binary relations. Observe that a binary relation defined on a set D is nothing other than a subset of D^2.

> We often denote binary relations by Greek letters in informal situations or we use the appropriate local notation, but we shall need a more systematic way of denoting them for the purposes of proofs. This will be explained later.

There is a nice graphical way to represent binary relations, at least when the sets they are defined on are not too big. Let ρ be a binary relation defined on the set D. We can draw a *directed graph* or *digraph* of ρ. This consists of *vertices*, labelled by the elements of D, and *arrows*: where an arrow is drawn from a to b precisely when $(a, b) \in \rho$.

---

[6]At least informally. There is the problem of Russell's paradox described in Example 2.1.3. How that can be dealt with is left to a more advanced course.

Example 3.1.2. The binary relation

$$\rho = \{(1,1),(1,2),(2,3),(4,1),(4,3),(4,5),(5,3)\}$$

is defined on the set $D = \{1,2,3,4,5\}$. Its corresponding directed graph is

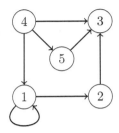

Example 3.1.3. Binary relations are very common in mathematics. Here are some examples.

1. The relation $x \mid y$ is defined on the set of natural numbers $\mathbb{N} = \{0,1,2,3,\ldots\}$ if x exactly divides y. For example, $2 \mid 4$ and $3 \mid 9$ but it is not true that $2 \mid 3$.

2. The relations $\leq$ (less than or equal to) and $<$ (strictly less than) are defined on $\mathbb{Z} = \{\ldots, -3, -2, -1, 0, 1, 2, 3, \ldots\}$.

3. The relation $\subseteq$ (is a subset of) is defined between sets.

4. The relation $\in$ (is an element of) is defined between sets.

5. The relation $\equiv$ (is logically equivalent to) is defined on the set of all wff in PL.

The set of all ordered n-tuples of elements of the set D is denoted by D^n. Subsets of D^n are called *n-ary relations*. As we have seen, for small values of n, there are particular terms that can be used to replace the generic term 'n-ary': when $n = 1$ we have *unary relations* (which are just subsets), when $n = 2$, we have *binary relations* and when $n = 3$ we have *ternary relations*. If ρ is a binary relation then it is a set of ordered pairs. Thus, we can ask whether $(a,b) \in \rho$ or not. However, it is more common in mathematics to write $a \rho b$ instead of $(a,b) \in \rho$. This is based on the usual way binary relations are written in mathematics. For example, on the set $\mathbb{N}$, there is the binary relation $\leq$. We would naturally write $a \leq b$ rather than $(a,b) \in \leq$.

It is common in mathematics to confuse a 1-place predicate with the subset it determines. Thus, the 1-place predicate 'x is mortal' gives rise to the set, let's call it M, of all mortal people. We could say 'Socrates is an element of M' but we are much more likely to say 'Socrates is mortal'. Similarly, 2-place predicates are often confused with the binary relations they define. Thus the 2-place predicate 'x is the father of y' determines a binary relation, say ϕ, where $(a, b) \in \phi$ precisely when 'a is the father of b', but it would be normal to refer to the binary relation 'x is the father of y'. As long as the precise mathematical definitions are understood, it should not be confusing to relax terminological niceties.

Functions can be (but don't have to be) regarded as relations. This will be discussed in Section 3.1.3.

3.1.3 Structures

Relations are organized into what are called structures. Before defining what these are, we begin with two examples.

Example 3.1.4. Here is part of the family tree of some Anglo-Saxon kings and queens of England:

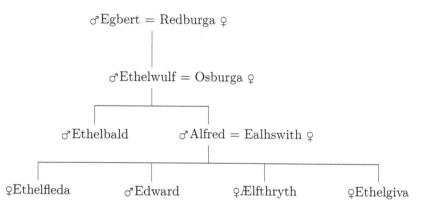

We shall now describe this family tree in mathematical terms. First, there is a set D of kings and queens called the domain. This consists of eleven Tolkienesque elements

$$D = \{\text{Egbert, Redburga, Ethelwulf, Osburga, Ethelbald, Alfred,} \\ \text{Ealhswith, Ethelfleda, Edward, Ælfthryth, Ethelgiva}\}.$$

The family tree contains more information than just a set of ancestors. Alongside each name is a symbol, ♂ or ♀, indicating whether that person is, respectively, male or female. This is just a way of defining two subsets of D:

$$M = \{\text{Egbert, Ethelwulf, Ethelbald, Alfred, Edward}\}$$

and

$F = \{\text{Redburga, Osburga, Ealhswith, Ethelfleda, Ælfthryth, Ethelgiva}\}.$

There are also two other pieces of information. The most obvious are the lines linking one generation to the next. This is the binary relation π defined by the 2-place predicate 'x is the parent of y'. For example,

$$(\text{Ethelwulf, Alfred}), (\text{Osburga, Alfred}) \in \pi.$$

A little less obvious is the notation $=$, which here is not the mathematical symbol for equality, but which stands for the binary relation μ defined by the 2-place predicate 'x is married to y'. For example,

$$(\text{Egbert, Redburga}), (\text{Redburga, Egbert}), (\text{Ethelwulf, Osburga}) \in \mu.$$

It follows that the information contained in the family tree is also contained in the following package

$$(D, M, F, \pi, \mu)$$

which is our first example of a structure (we shall formally define this term below).

Example 3.1.5. The package

$$(\mathbb{N}, \mathbb{E}, \mathbb{O}, \leq, |)$$

consists of the set of natural numbers, $\mathbb{N}$, two subsets $\mathbb{E}$ and $\mathbb{O}$, being respectively the sets of odd and even natural numbers, and two binary relations $\leq$ and $|$. At first sight, this looks quite different from the family tree in Example 3.1.4 but is, in fact, mathematically close kin since it also consists of a non-empty set, two subsets and two binary relations.

In general, we define a *structure* D to consist of a non-empty set D, called the *domain*[7] of the structure, together with a list of relations of various arities — thus a list of subsets of D, a list of binary relations on D, and so on — rounded off with a list of distinguished elements of D; we require at least one relation but the presence of distinguished elements is not obligatory. Schematically, a structure has the following form:

$$\mathsf{D} = (D, \text{relations, distinguished elements}).$$

Both Example 3.1.4 and Example 3.1.5 are structures. Although structures usually have a finite number of relations or distinguished elements there are, in fact, no restrictions on how many of either there might be. However, the case where there are infinitely many distinguished elements arises mainly for what might be called 'technical reasons', which will be explained later.

The extra ingredient not discussed in the examples above is the presence of distinguished elements. Such elements are very common in mathematics.

---

[7]This is a different use of the word 'domain' from that which occurs in the definition of a function. If mathematicians aren't using a multiplicity of words to describe the same thing, they are using one word to mean many things. Context, as always, is important.

In first-order logic, structures are usually defined in a slightly more general way to be the following:

$$\mathsf{D} = (D, \text{relations}, \text{functions}, \text{distinguished elements})$$

where functions are also included. Now, in fact, functions can be represented by means of relations. Thus a function $f\colon D \to D$ can be viewed as the binary relation $\{(a, f(a))\colon a \in D\}$ and a function $f\colon D^2 \to D$ can be viewed as the ternary relation $\{(a, b, f(a, b))\colon (a, b) \in D^2\}$. So, by restricting to structures where there are no functions we do not lose any expressive power. However, we do lose convenience and a certain naturalness of expression so that in more advanced work functions are always explicitly included.

Examples 3.1.6.

1. Trees are structures, being particular kinds of binary relations, but we always want to refer to the root of the tree: this is a distinguished vertex.

2. Boolean algebras are structures. Recall that a Boolean algebra is defined to be a structure of the following form $(B, +, \cdot, ^-, 0, 1)$ where $+$ and $\cdot$ are functions from B^2 to B, where $^-$ is a function from B to B and where 0 and 1 are distinguished elements.

Looking ahead a little, FOL is a language that will enable us to talk about structures.

3.1.4 Quantification

We begin with an example. Let $M(x)$ be the 1-place predicate 'x is mortal'. I want to say that 'everything is mortal'. In PL, I could try to use infinitely many conjunctions

$$M(\text{Socrates}) \wedge M(\text{Darth Vader}) \wedge M(\text{Winnie-the-Pooh}) \wedge \dots$$

but this conceals problems about different kinds of infinities and structures of different sizes; in short, this won't work. We get around these problems by introducing something new: what is called the *universal quantifier* $\forall$. We write

$$(\forall x)M(x)$$

which should be read 'for all x, x is mortal' or 'for each x, x is mortal'. The variable does not have to be x, we have all of these $(\forall y)$, $(\forall z)$ and so on. As we shall see, you should think of $(\forall x)$, and its ilk, as being a new unary connective. Thus

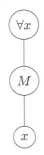

is the parse tree for $(\forall x)M(x)$.

The corresponding infinite disjunction

$$M(a) \lor M(b) \lor M(c) \lor \dots$$

is true when at least one of the terms is true. There is a corresponding *existential quantifier* $\exists$. We write

$$(\exists x)M(x)$$

which should be read 'for some x, x is mortal' or 'there exists an x, such that x is mortal' or 'there is at least one x, such that x is mortal'. The variable does not have to be x, of course, we have all of these $(\exists y)$, $(\exists z)$ and so on. You should also think of $(\exists x)$ and its ilk as being a new unary connective. Thus

is the parse tree for $(\exists x)M(x)$.

> In English, we are forced to choose between saying 'everything' or 'everyone'. Natural languages are full of such distinctions — animate versus inanimate, human versus non-human, whether someone or something is masculine, feminine or neuter, etc. — which play no role in logic. Just as with grammatical forms — such as 'is' versus 'are', for example — it is usual to make silent adjustments to obtain natural English.

3.1.5 Syntax

In the last footnote, I warned the reader about the way that words in mathematics can often have different meanings in different contexts. This is true for our next definition. I defined the word 'language' towards the end of Section 1.2. Now that same word will be used with a different meaning. In fact, it is really being used in the sense of 'alphabet', also defined at the end of Section 1.2.

A *(first-order) language L* consists of the following ingredients:

- A list of *constants* $a_1, a_2, a_3, \ldots$ where we allow for either no constants a finite number of constants, or infinitely many constants. In practice, constants will be denoted by $a, b, c, \ldots$. For 'technical reasons', we shall also allow some of the constants to be drawn from a set U. This will be explained later.

- A list of *variables* $x_1, x_2, x_3, \ldots$. In practice, variables will be denoted by $x, y, z, \ldots$. A constant or variable is called a *term*.[8]

- A (non-empty) list of *predicate symbols*; a predicate symbol is simply a letter, such as $A, B, C, \ldots$, and a finite natural number $n \geq 1$, called its *arity*, associated with that letter. If A has arity n we shall usually say that A is an *n-place predicate symbol*. The arity of a predicate symbol is usually indicated by the number of terms that are allowed to follow it. Thus if we write $A(x)$ we know that A is a 1-place predicate symbol, whereas if we write $A(x, y)$ we know that A is a 2-place predicate symbol.

- All *propositional connectives*: $\neg, \wedge, \vee, \rightarrow, \leftrightarrow, \oplus$.

- *Quantifiers* $(\forall x_1), (\forall x_2) \ldots$ and $(\exists x_1), (\exists x_2) \ldots$, though we shall frequently use $(\forall x), (\forall y) \ldots$ and $(\exists x), (\exists y) \ldots$.

  A language, as defined above, provides the linguistic resources to talk about something (actually, to talk about a structure). The constants and predicate symbols required will be determined by the application in mind.

As we have seen, an *atomic formula* is an expression of the form $A(y_1, \ldots, y_n)$ where A is a predicate symbol of arity n and each y_i is a term for $1 \leq i \leq n$. More generally, a *formula* in FOL is defined in the following cumulative way:

(F1) All atomic formulae are formulae.

(F2) If X and Y are formulae so too are $(\neg X)$, $(X \wedge Y)$, $(X \vee Y)$, $(X \rightarrow Y)$, $(X \leftrightarrow Y)$, $(X \oplus Y)$, and $(\forall x)X$, for any variable x, and $(\exists x)X$, for any variable x.

---

[8]The notion of term becomes more interesting, and more complex, when we allow function symbols.

(F3) All formulae arise by repeated application of steps (F1) or (F2) a finite number of times.

I will carry forward from PL to FOL, without further comment, the same conventions concerning the use of brackets. We may also adapt parse trees to FOL, as we have already intimated.

Example 3.1.7. The formula $(\forall x)[(P(x) \to Q(x)) \wedge S(x,y)]$ has the following parse tree:

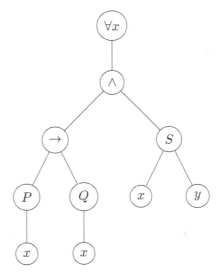

It is important to remember that a language is just a list of symbols. For example, the predicate symbols have no meaning in themselves, nor do the formulae. When we define the semantics of a first-order language these meaningless ingredients will acquire meaning.

Expansion of a language by adjoining constants

A language L is chosen so that it has the resources to 'talk' about a structure; it is this which determines which predicate symbols, along with their arities, and which constants we need. However, for certain 'technical reasons' we might want to extend a language by adding new constants. An *expansion* of the language L is a language M which has the same predicate symbols but the list of constants of L is extended by adjoining a list of new constants; if those new constants are chosen from the set U then we say that we have *adjoined constants from U*. If M is obtained from L by adjoining just one new constant, we say that M is a *1-step expansion* of L.

3.1.6 Variables: their care and maintenance

This section is technical, but vitally important for understanding quantifiers. It contains some heavy-duty definitions whose significance will only become clear once we fully develop FOL.

Subtree and subformula

Choose any vertex v in a tree, except a leaf. The whole part of the tree, including v itself, that lies below v is clearly also a tree. It is called the *subtree determined by v*. We illustrate this definition below.

Example 3.1.8. Consider the following tree:

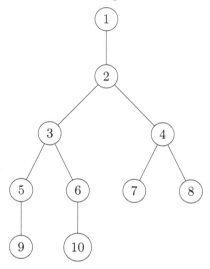

The subtree determined by the vertex 4 is the following tree:

whereas the subtree determined by the vertex 3 is the following tree:

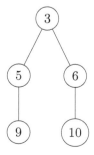

In a parse tree, a subtree determines a *subformula* which is the formula obtained by regarding the subtree as a parse tree in its own right. Observe that in defining a subformula, the lowest level we go down to is that of an atomic formula.

Example 3.1.9. Let's return to Example 3.1.7. The subtree determined by $\rightarrow$ is the following:

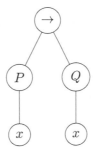

This is the parse tree for the formula $P(x) \rightarrow Q(x)$ which is therefore a subformula. The subtree determined by $\wedge$ is the following:

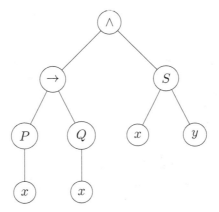

This is the parse tree for the formula $(P(x) \rightarrow Q(x)) \wedge S(x, y)$ which is therefore a subformula.

Occurrence of a variable

In a formula, a given variable may occur a number of times.

Example 3.1.10. Consider the formula $(P(x) \rightarrow Q(y)) \rightarrow (\exists y)R(x, y)$. Then the variable x occurs twice $(P(\boldsymbol{x}) \rightarrow Q(y)) \rightarrow (\exists y)R(\boldsymbol{x}, y)$ and the variable y occurs thrice $(P(x) \rightarrow Q(\boldsymbol{y})) \rightarrow (\exists \boldsymbol{y})R(x, \boldsymbol{y})$.

We can therefore talk about the *occurrences* of a variable. Each occurrence of a given variable in a formula will be in a different environment. We can also talk about *occurrences of quantifiers* in much the same way.

Bound and free variables

We now come to a key definition in our description of FOL. This tells us that variables can play two different roles in a formula. These two different roles are determined solely by syntax but they influence the semantics, that is the meaning, of the formula. First of all, an occurrence of a quantifier such as $(\forall x)$ or $(\exists x)$ determines a subformula, called the *scope* of that quantifier.[9] Any occurrence of the variable x in that subformula is said to be *bound*.[10] Any occurrence of the variable x that is not in a subformula determined by some $(\forall x)$ or $(\exists x)$ is said to be *free*. To be clear: all occurrences of variables x in the quantifiers $(\forall x)$ and $(\exists x)$ are also bound. In addition, a variable may have both free and bound occurrences in a formula and so the two notions are not mutually exclusive.

Example 3.1.11. Consider the formula $(P(x) \to Q(y)) \to (\exists y)R(x,y)$. I shall highlight each of the variables in turn and say whether it is free or bound. The first occurrence of the variable x in

$$(P(\boldsymbol{x}) \to Q(y)) \to (\exists y)R(x,y)$$

is free since x is not in the scope of either $(\forall x)$ or $(\exists x)$. Next, we have an occurrence of y in

$$(P(x) \to Q(\boldsymbol{y})) \to (\exists y)R(x,y)$$

which is also free. However, the next occurrence of y in

$$(P(x) \to Q(y)) \to (\exists \boldsymbol{y})R(x,y)$$

is bound since it occurs within the scope of $(\exists y)$. The next occurrence of x in

$$(P(x) \to Q(y)) \to (\exists y)R(\boldsymbol{x},y)$$

is clearly free. Finally, the last occurrence of y in

$$(P(x) \to Q(y)) \to (\exists y)R(x,\boldsymbol{y})$$

is bound since it occurs in the subformula $(\exists y)R(x,y)$. We may summarize by saying that the variable x occurs free twice, and the variable y occurs free once and bound twice.

A formula that has a free occurrence of some variable is said to be *open*, otherwise it is said to be *closed*. A closed formula is called a *sentence*. There is one feature of the definition of a sentence that I should highlight: it may well have constants. The definition of a sentence only requires that there be no free variables.

---

[9]The scope of a quantifier is therefore the largest subformula over which that quantifier holds dominion.

[10]The word 'bound' is an adjective. The corresponding verb is 'to bind'.

Example 3.1.12. The formula $(\forall x)R(x, a)$ is a sentence since both occurrences of the variable x are bound.

> This usage of the word 'sentence' conflicts with how we used the word 'sentence' back in Section 1.1. In FOL, a sentence is, by definition, a closed formula.

In terms of parse trees, a formula is a sentence if there is a path from each variable x (regarded as a leaf) to a vertex labelled $\forall x$ or $\exists x$.

Example 3.1.13. The formula

$$(\exists x)(F(x) \wedge G(x)) \to ((\exists x)F(x) \wedge (\exists x)G(x))$$

is a sentence since all occurrences of the variable x are bound.

> **Notation.** If X has free variables $x_1, \ldots, x_n$, where $n \geq 1$, then we shall write $X[x_1, \ldots, x_n]$ using square brackets.

Substitution

The significance of bound and free variables is inextricably linked with a process called *substitution*. We begin with a familiar example.

Example 3.1.14. In the UK, many words are spelt differently from in the US. Thus, in the UK we write 'colour' whereas in the US this is written 'color'. A British author will write a book using UK spellings, but his US publisher might require him to use US spellings.[11] This is easy enough to deal with: the author simply uses a word processing program to replace all occurrences of the word 'colour', for example, by the word 'color'; that is, we *substitute* the word 'color' for the word 'colour' wherever it appears.

Substitution in logic is essentially the same 'search and replace' operation as in word processing. However, there is one important caveat: whereas free variables are replaced, bound variables are not: they are out-of-bounds. I now make this idea precise. Let X be any formula and let $*$ be a term; that is, either a variable or a constant. We define an operation $(x \leftarrow *)$, which should be read '*replace x by $*$*' or '*$*$ is substituted for x*', with argument a formula X, as follows:

- If x does not occur free in X then $(x \leftarrow *)(X) = X$.

- If x does occur free in X then the formula $(x \leftarrow *)(X)$ is obtained from X by replacing all *free* occurrences of x in X by $*$.

If x occurs freely in the formula X then we write $X[x]$ and it is then natural to write $(x \leftarrow *)(X[x]) = X[*]$ for the result of replacing all free occurrences of x in X by $*$. Substitution by a variable is also called *change of variable*.

---

[11] The publisher of this book has been exemplary in tolerating my orthographical idiosyncrasies.

Examples 3.1.15. Here are two concrete examples of substitution. If $X = P(x) \wedge (\exists x)Q(x)$ then

$$(x \leftarrow a)(P(x) \wedge (\exists x)Q(x)) = P(a) \wedge (\exists x)Q(x).$$

If $X = P(x) \wedge (\forall y)Q(x, y)$ then

$$(x \leftarrow z)(P(x) \wedge (\forall y)Q(x, y)) = P(z) \wedge (\forall y)Q(z, y).$$

3.1.7 Semantics

In Section 3.1.5, I defined the syntax of FOL and in Section 3.1.6, I defined some syntactic terms. We now turn to the meaning of FOL or, more formally, its semantics. In principle, this is simple but in practice there is a lot of bureaucracy to deal with that can obscure that underlying simplicity: *bon courage.*

Interpreting languages

Let L be a first-order language. This means that we have selected a list of predicate symbols of various arities and a list of constants (possibly none); recall that variables and propositional connectives come for free. For example, our language might consist of two predicate symbols P and Q, where P is a 1-place predicate symbol and Q is a 2-place predicate symbol, and a single constant a. Our language would therefore be (P, Q, a). Using this language we can construct formulae but so far these have no meaning; they are simply strings of symbols. They spring into life only when we interpret them. An *interpretation* I of a first-order language L is a package that consists of the following ingredients:

- A structure D, which consists of a set D, the domain, and a list of n-ary relations and a list of distinguished elements of D.

- An assignment of a specific n-ary relation R^I to each n-place predicate symbol R in the language. To avoid superscripts, I shall usually write $\mathcal{R}$ instead of R^I in proofs as long as the interpretation is clear.[12] Where the relations in question are well-known, I shall apply whatever notation is usual such as $\leq$ or ρ, etc.

- An assignment of a distinguished element a^I of D to each constant a in the language. To avoid superscripts, I shall usually write d, or a similar element, instead of a^I.

An interpretation is said to be *finite* if it has a finite domain, otherwise it is said to be *infinite*.

---

[12]The salient feature of this notation is that R and $\mathcal{R}$ are the same letter in different fonts.

Example 3.1.16. Consider the language (P, Q, a), where P is a 1-place predicate symbol, Q is a 2-place predicate symbol and a is a constant. An interpretation of this language would therefore be a structure of the following form

$$\mathsf{D} = (D, \mathcal{P}, \mathcal{Q}, d),$$

where D, the domain is a non-empty set, $\mathcal{P}$ is a subset of D, $\mathcal{Q}$ is a binary relation on D, and d is an element of D, and where we interpret P as $\mathcal{P}$, Q as $\mathcal{Q}$ and a as d.

> The basis of the definition of an interpretation is the most primitive idea of how languages work: the meaning of a word is the thing it refers to.

Interpreting formulae

Once we have an interpretation of the symbols of the language, we can interpret formulae. Before we describe how this is done, we give some examples to establish the basic idea.

Example 3.1.17. We shall interpret the 2-place predicate symbol $P(x, y)$ as the binary relation $\mathcal{P}$ determined by the expression 'x is the father of y' where the domain D is the set of people. Observe that the expression 'x is the father of y' is neither true nor false since we know nothing about x and y. The binary relation $\mathcal{P}$ is defined by $(a, b) \in \mathcal{P}$ if and only if 'a is the father of b' is true; recall that in practice we would often confuse the binary relation with the 2-place predicate that defines it, whereas here I am being formally precise. Consider now the formula $(\exists y)P(x, y)$. This gives rise to the expression 'x is a father' which is a 1-place predicate. The subset it determines is

$$\{a : a \in D \text{ and } a \text{ is a father}\}.$$

Finally, consider the formula $S = (\forall x)(\exists y)P(x, y)$. This gives rise to the expression that says 'everyone is a father'. Observe that S is a sentence and that 'everyone is a father' is a statement. In this case, it is false.

Example 3.1.18. We shall interpret the 2-place predicate symbol $P(x, y)$ as the binary relation $\mathcal{P}$ defined on the set $\mathbb{N} = \{0, 1, 2, \ldots\}$ of natural numbers by the 2-place predicate $x \leq y$. The binary relation $\mathcal{P}$ is defined by $(a, b) \in \mathcal{P}$ if and only if $a \leq b$. The formula $(\exists y)P(x, y)$ gives rise to the expression $(\exists y)(x \leq y)$ which is a 1-place predicate. The subset it determines is

$$\{a : a \in \mathbb{N} \text{ and } a \leq b \text{ for some } b \in \mathbb{N}\}$$

which a simple check shows is the whole set $\mathbb{N}$. Finally, the sentence $S = (\forall x)(\exists y)P(x, y)$ gives rise to the expression 'there is no largest natural number' which is true.

Example 3.1.19. Consider the language with two 1-place predicate symbols A and B and two 2-place predicate symbols P and Q. We shall interpret formulae of this language in the structure which is the family tree of Example 3.1.4. Thus the domain D is the given set of Anglo-Saxon kings and queens; A is interpreted as the subset M of males; B is interpreted as the subset F of females; P is interpreted as the relation π 'is a parent of'; and Q is interpreted as the relation μ 'is married to'. We now interpret some formulae in this language. Consider, first, the formula $W_1 = (\exists z)(P(x, z) \wedge P(z, y))$. This translates into the expression $(\exists z)((x \, \pi \, z) \wedge (z \, \pi \, y))$ which is a 2-place predicate. The binary relation corresponding to it is $a \, \gamma \, b$ defined precisely when $(\exists c)((a \, \pi \, c) \wedge (c \, \pi \, b))$ is true but which we would describe in English by the phrase 'a is the grandparent of b'. Now consider $W_2 = (\exists z)(P(x, z) \wedge P(z, y)) \wedge A(x)$, a slight modification of W_1. Then W_2 is interpreted as the binary relation $a \, \phi \, b$ defined precisely when 'a is the grandfather of b'. Finally, consider the formula $W_3 = (\exists z)(P(z, x) \wedge P(z, y))$. This translates into the expression $(\exists z)((z \, \pi \, x) \wedge (z \, \pi \, y))$ whose corresponding binary relation is determined by the phrase 'a and b are siblings'. These examples demonstrate that our basic language enables us to describe kinship relations amongst the Anglo-Saxon nobility.

We see from these examples that a sentence will either be true or false in an interpretation; we shall say that the sentence has a *truth value* in an interpretation and that truth value is its interpretation. A formula with n free variables will describe an n-ary relation in an interpretation by means of an expression which is a predicate of arity n; the n-ary relation is the interpretation of the formula.

Keeping the above examples in mind, we now define the meaning to be ascribed to a formula in an interpretation. The first step is to define precisely what we mean by an 'expression'. Intuitively, this is nothing other than an n-ary predicate constructed from the predicates and distinguished elements in the interpretation using the logical connectives and quantifiers. Let I be an interpretation of the language L in the structure D. An *expression* in this interpretation is defined in the following way:

(E1) For each n-place predicate symbol P, we say that $((y_1, \dots, y_n) \in \mathcal{P})$ is an *expression* where each y_i is either a variable or a distinguished element of D. The outer brackets can be omitted when there is no danger of ambiguity.

(E2) If X and Y are expressions so too are $(\neg X)$, $(X \wedge Y)$, $(X \vee Y)$, $(X \to Y)$, $(X \leftrightarrow Y)$, $(X \oplus Y)$ and $(\forall x)X$, for any variable x, and $(\exists x)X$, for any variable x.

(E3) All expressions arise by repeated application of steps (E1) or (E2) a finite number of times.

An expression is a formula with (some minor typographical changes) in a language that uses actual predicates as predicate symbols and distinguished elements as constants. This means that it should be clear what we mean by talking about the free and bound variables of an expression.

Next, we describe how to use a formula X in a language L as a blueprint to construct an expression $I(X)$ where I is an interpretation of L in a structure D.

- If $X = P(y_1, \ldots, y_n)$, where each y_i is either a variable or a constant, then $I(X) = ((z_1, \ldots, z_n) \in \mathcal{P})$ where $z_i = y_i$ if y_i is a variable otherwise z_i is the interpretation of the constant y_i.

- If $X = \neg Y$ then $I(X) = \neg I(Y)$; if $X = Y \bullet Z$, where $\bullet$ is any of the binary logical connectives, then $I(X) = I(Y) \bullet I(Z)$.

- If $X = (\forall x)Y$ then $I(X) = (\forall x)I(Y)$ and if $X = (\exists x)Y$ then $I(X) = (\exists x)I(Y)$.

This definition of the construction of an expression from a formula continues with the primitive idea of how languages work: we simply translate X to $I(X)$ by consulting the 'dictionary' I.

Example 3.1.20. Consider the following sentence

$$(\exists x)(\forall y)P(x, y) \wedge (\exists z)(A(z) \vee B(z)).$$

Its corresponding expression would be

$$(\exists x)(\forall y)((x, y) \in \mathcal{P}) \wedge (\exists z)((z \in \mathcal{A}) \vee (z \in \mathcal{B})).$$

We now describe how to determine the interpretation of a formula. There are two cases: the first, where the formula is a sentence and the second, where the formula has free variables.

1. If X is a sentence then $I(X)$ will make an assertion about the structure D that will either be true or false. Thus $I(X)$ has a truth value which is its interpretation. We show how to determine the truth value of $I(X)$ in the interpretation:

   (a) If $I(X)$ is directly constructed from two expressions using the propositional connectives then their usual meaning is carried forward. For example, if $I(X)$ is equal to $I(X_1) \wedge I(X_2)$ then $I(X)$ is true in the interpretation precisely when $I(X_1)$ and $I(X_2)$ are both true in the interpretation. In other words, we just use the meanings ascribed to the propositional connectives in Chapter 1.

(b) Suppose that $X = (\forall x)X_1$. If x does not occur free in X_1 then the truth value of $I(X)$ is the same as the truth value of $I(X_1)$. If x does occur free in X_1 then $I(X)$ is true precisely when the truth value of $I(X)[d]$ is true for **all** elements $d \in D$.

(c) Suppose that $X = (\exists x)X_1$. If x does not occur free in X_1 then the truth value of $I(X)$ is the same as the truth value of $I(X_1)$. If x does occur free in X_1 then $I(X)$ is true precisely when the truth value of $I(X)[d]$ is true for **some** element $d \in D$.

2. Suppose now that X has free variables $x_1, \ldots, x_n$ where $n \geq 1$. Thus we can write $X = X[x_1, \ldots, x_n]$. The meaning of X will be an n-ary relation $\mathcal{X}$ on the set D. To define this relation, we need to determine when $(d_1, \ldots, d_n) \in \mathcal{X}$ where $d_1, \ldots, d_n \in D$. It is at this point that the 'technical reasons' alluded to earlier come into play and where it is convenient to use the notion of adjoining constants to a language. We adjoin $d_1, \ldots, d_n$ to the language L to obtain the language L' and we extend the interpretation I to the interpretation I' where we interpret each d_i by itself. We now substitute d_i for x_i in X for $1 \leq i \leq n$ to obtain the *closed* formula $X[d_1, \ldots, d_n]$ (that is, closed in the extended language). We may now define $(d_1, \ldots, d_n) \in \mathcal{X}$ precisely when $X[d_1, \ldots, d_n]$ is true in the extended interpretation; we know how to determine this by case (1) above.

As I warned at the beginning of this section, there is a lot of bureaucracy involved in the above definitions. Here is a summary of what we have done so far:

A formula X constructed from a language L should be regarded as a blueprint. Once an interpretation I (with underlying structure D) of the language L has been chosen, the blueprint X is used to construct a concrete expression $I(X)$. If X has n free variables then the expression $I(X)$ will be an n-place predicate and so will describe a subset $\mathcal{X}$ of D^n and therefore an n-ary relation $\mathcal{X}$ on D; the relation $\mathcal{X}$ is the interpretation of X. If X has no free variables $I(X)$ will be a statement about the structure D that will have a truth value; that truth value is the interpretation of X. Finally, it is important to realize that the notion of semantics we have described is philosophically simple-minded.

Universally valid formulae

Let S be a sentence. We say that I *is a model of S*, written $I \vDash S$, if S is true in the interpretation I. When the interpretation is understood, we shall also say that a structure is a model of a sentence. We say that such a structure is a model of a set of sentences $\mathcal{B}$ if it is a model of every sentence in $\mathcal{B}$. If a sentence S has a model then it is said to be *satisfiable*. An interpretation of a sentence S in which that sentence is false is called a *counterexample*. A sentence S is said to be *universally valid*, written ⊨ S, if it is true in **all** interpretations.

Let X be a formula with free variables $x_1, \ldots, x_n$. Then the *universal closure* of X is the sentence $(\forall x_1) \ldots (\forall x_n)X$. We define X to be universally valid if the universal closure of X is universally valid.[13] Let S_1 and S_2 be formulae. We write $S_1 \equiv S_2$, and say that S_1 and S_2 are *logically equivalent*, if ⊨ $S_1 \leftrightarrow S_2$. If S_1 and S_2 are sentences then to say that they are logically equivalent is to say that in every interpretation they have the same truth value. If S_1 and S_2 are formulae with the same free variables then to say that they are logically equivalent is to say that they define the same relations in every interpretation.

We can also generalize our definition of what constitutes a valid argument from Section 1.10. If $A_1, \ldots, A_n, B$ are sentences we write $A_1, \ldots, A_n \models B$ to mean that if I is any interpretation in which all of $A_1, \ldots, A_n$ are true then B must also be true in that interpretation. It is not hard to prove that this is equivalent to showing that ⊨ $(A_1 \wedge \ldots \wedge A_n) \to B$. Thus we can now determine when arguments of the form $A_1, \ldots, A_n \therefore B$ are valid where the formulae appearing are from first-order logic.

> Whereas in PL we studied tautologies, in FOL we study universally valid formulae, especially universally valid sentences.

Renaming bound variables

What I say about universal quantifiers below applies equally well to existential quantifiers

The grammar of FOL forces us to choose variables but not all choices lead to different meanings. We begin with two examples to illustrate what we mean by this.

Example 3.1.21. Consider the sentence $(\forall x)A(x)$. This is different from the sentence $(\forall y)A(y)$ which, in turn, is different from the sentence $(\forall z)A(z)$. Moreover, all the sentences $(\forall x_i)A(x_i)$, where $i \in \mathbb{N}$, are different from each other and the three sentences above. *Syntactically*, all of these sentences are distinct. Despite this, in every interpretation, these sentences will all say the same thing; that is, they have the same meaning. This is because the sentence $(\forall x)A(x)$ is true in a structure $(D, \mathcal{A})$ when for every $d \in D$, we have that $d \in \mathcal{A}$. The phrase 'for every $d \in D$, we have that $d \in \mathcal{A}$' makes no mention of the variable x and so it follows that

$$(\forall x)A(x) \equiv (\forall y)A(y) \equiv (\forall z)A(z) \equiv (\forall x_i)A(x_i).$$

---

[13]The definition of the universal closure of a formula is not unique because it depends on the order of the variables. We shall prove later, however, that if all the quantifiers are of the same type — that is, all universal quantifiers or all existential quantifiers — then the order in which they appear doesn't matter.

All of these sentences have the same shape $(\forall *)A(*)$, where $*$ is any variable, but we are forced by the syntax to choose a specific variable to replace $*$ even though what we choose doesn't matter from the point of view of semantics.

Example 3.1.22. Consider now the formula $(\forall x)B(x,y)$. Here x is bound and y is free. In the light of what we described above, the meaning of $(\forall x)B(x,y)$ and the meaning of $(\forall z)B(z,y)$ should be the same in any interpretation. Let D be a structure with domain D and binary relation $\mathcal{B}$ where B is interpreted as $\mathcal{B}$. Then the interpretation of $(\forall x)B(x,y)$ is the subset of D consisting of all those elements b such that $(a,b) \in \mathcal{B}$ for all $a \in D$. Once again, there is no mention of the variable x in the interpretation and so this subset is also the interpretation of the formula $(\forall z)B(z,y)$. Thus $(\forall x)B(x,y)$ and $(\forall z)B(z,y)$ have the same meanings. But here we have to be careful. What about the formula $(\forall y)B(y,y)$? Now I am in trouble, because this has no free variables and so cannot possibly have the same meaning as $(\forall x)B(x,y)$ which does have a free variable. What has gone wrong? The problem is that the variable y is free in $(\forall x)B(x,y)$ but becomes bound in $(\forall y)B(y,y)$. *This is a syntactic warning that the meaning has been changed.* It follows that I can replace the $*$ in the expression $(\forall *)B(*,y)$ by any variable *except* y and obtain a formula that will all give rise to the same set in any interpretation.

These two examples raise the question of when we can and when we can't change bound variables in a formula without changing the meaning of that formula in any interpretation. To resolve this question, we need the following definition. Let W be any formula. We say that y *is free (to be substituted) for* x in W if no free occurrence of x in W occurs within the scope of either $(\forall y)$ or $(\exists y)$ in W.

Lemma 3.1.23 (Renaming bound variables). *Let X be a formula and let y be a variable that does not occur free in X. If y is free for x in X then $(\forall x)X$ and $(\forall y)(x \leftarrow y)X$ will always have the same interpretations.*

Proof. Let X have the free variables $x, x_1, \ldots, x_n$. By assumption, y is not equal to any of these variables. We now determine the free variables in $(x \leftarrow y)X$ and their positions. What might go wrong is that a free occurrence of x might be in the scope of some occurrence of $(\forall y)$ or $(\exists y)$. This cannot happen by assumption. Thus the free variables in $Y = (x \leftarrow y)X$ are precisely $y, x_1, \ldots, x_n$, with the added property that where x is free in X then y is free in Y (and vice versa), and where x_i is free in X then x_i is free in Y (and vice versa). It is now clear that if $(\forall x)X$ and $(\forall y)Y$ are sentences then they are true or false in the same interpretations, and if $(\forall x)X$ and $(\forall y)Y$ have free variables then they determine the same relations in the same interpretations. $\square$

We often rename bound variables within subformulae of a formula so that they are distinct from free variables. This is useful in some applications but can also help in understanding what a formula is saying.

Example 3.1.24. Consider the formula $(\forall x)P(x,y) \wedge Q(x)$. The variable x is bound in $(\forall x)P(x,y)$ and free in $Q(x)$. I can rename the bound occurrences of x in the subformula $(\forall x)P(x,y)$ by z, since z is not a free variable in $(\forall x)P(x,y)$ and z is free for x in $P(x,y)$, to obtain the formula $(\forall z)P(z,y) \wedge Q(x)$ where now the variable x is free wherever it occurs.

> The fact that different formulae have the same meaning, differing only in the inessential choice of variables, is analogous to the way that one and the same fraction may appear in different guises: thus $\frac{1}{2}$ and $\frac{2}{4}$ look different but actually name the same rational number. Similarly, just as adding two fractions often requires the fractions be renamed (by making the denominators the same) so too there are results in logic that require the variables in formulae be renamed. A good example occurs in the proof of the result (not included in this book) that every formula is equivalent to one in which all quantifiers appear at the front; a formula that has this shape is said to be in *prenex normal form*.

3.1.8 De Morgan's laws for quantifiers

We have seen that the universal quantifier $\forall$ is a sort of infinite version of $\wedge$ and the existential quantifier is a sort of infinite version of $\vee$. The following result is therefore not surprising. It is worth considering first the following example which contains the idea of the proof of part (1).

Example 3.1.25. Consider the statement 'not all swans are white'. How would you prove it? For Australian readers, this is easy: look out of the window and locate a black swan. That is, you establish that there exists a swan that is not white.

Theorem 3.1.26 (De Morgan for quantifiers). *Assume that $\neg(\forall x)X$ and $\neg(\exists x)X$ are sentences.*

1. $\neg(\forall x)X \equiv (\exists x)\neg X$.

2. $\neg(\exists x)X \equiv (\forall x)\neg X$.

Proof. (1) Suppose first that x is not free in X. Then $(\forall x)X \equiv X$ and $(\exists x)\neg X \equiv \neg X$. Thus the result is trivial. We may therefore assume in what follows that x is free in X. Our goal is to prove that in every interpretation where $\neg(\forall x)X$ is true then $(\exists x)\neg X$ is true and vice versa. Suppose that $\neg(\forall x)X$ is true in some interpretation where $X[x]$ is interpreted as the subset $\mathcal{X}$. Then $(\forall x)X$ is false in this interpretation. Thus there must be some element a of the domain of the interpretation such that $a \notin \mathcal{X}$ is true. Thus $(\exists x)\neg X$ is true in that interpretation. The other direction is proved in a similar way.

(2) By part (1) above, we have that $\neg(\forall x)\neg X \equiv (\exists x)\neg\neg X \equiv (\exists x)X$. Negating both sides of this logical equivalence gives the result. $\qquad\square$

3.1.9 Quantifier examples

This section is a collection of examples of formulae in FOL and how they should be interpreted. Quantifiers are the distinctive new ingredient in FOL and contribute much to its complexity, so our examples will revolve around quantifiers.

Example 3.1.27. When we use the universal quantifier, we often don't want to say 'all things have a property' but instead 'all things of a certain type have a property'. For example, I don't want to say that 'everyone likes carrots' (which is clearly false) but rather that 'all rabbits like carrots' (which is morally true). To formalize this in FOL, we need the 1-place predicate symbol $A(x)$ to be interpreted as the 1-place predicate 'x is a rabbit' and the 1-place predicate symbol $B(x)$ to be interpreted as the 1-place predicate 'x likes carrots'. The statement 'all rabbits like carrots' is now formalized by the following sentence:

$$(\forall x)(A(x) \to B(x)).$$

Let's see why. Flopsy is a rabbit. Thus $A(\text{Flopsy})$ is true. In our sentence above, we may substitute both occurrences of the free variable x in $A(x) \to B(x)$ by the name Flopsy to obtain $A(\text{Flopsy}) \to B(\text{Flopsy})$. But from $A(\text{Flopsy})$ is true and the fact that $A(\text{Flopsy}) \to B(\text{Flopsy})$ is true, it follows that $B(\text{Flopsy})$ is true and so Flopsy likes carrots. Flopsy was, of course, an arbitrary rabbit so the sentence does, indeed, say 'all rabbits like carrots'.

I now want to say 'some rabbits don't like carrots'. This means that there is at least one entity that is simultaneously a rabbit and doesn't like carrots.[14] The correct formalization of this sentiment in FOL is therefore

$$(\exists x)(A(x) \wedge \neg B(x)).$$

We can see that our two translations make sense since

$$\neg(\forall x)(A(x) \to B(x))$$

is logically equivalent to

$$(\exists x)(A(x) \wedge \neg B(x))$$

as the following calculation shows

$$\neg(\forall x)(A(x) \to B(x)) \equiv (\exists x)\neg(A(x) \to B(x)) \equiv (\exists x)\neg(\neg A(x) \vee B(x))$$

which is logically equivalent to $(\exists x)(A(x) \wedge \neg B(x))$, as claimed.

---

[14]Mopsy, perhaps.

Example 3.1.28. The use and abuse of quantifiers is a source of confusion in mathematics. If I write

$$(x + y)^2 = x^2 + 2xy + y^2,$$

for example, what I really mean is that the sentence

$$(\forall x)(\forall y)[(x + y)^2 = x^2 + 2xy + y^2]$$

is true when interpreted in the real numbers. On the other hand, when I ask you to solve

$$x^2 + 2x + 1 = 0$$

I am actually asking whether the sentence

$$(\exists x)[x^2 + 2x + 1 = 0]$$

is true when interpreted in the real numbers. By the way, I am using [and] here simply as brackets, nothing more.

Example 3.1.29. First year students of mathematics receive their first dose of mathematical culture shock when confronted with the formal definition of a continuous function. This runs as follows: the real-valued function f is continuous at the point x_0 if for all $\epsilon > 0$ there exists $\delta > 0$ such that for all x where $x_0 - \delta < x < x_0 + \delta$ we have that $f(x_0) - \epsilon < f(x) < f(x_0) + \epsilon$. One of the things that makes this definition hard to understand is that there are three quantifiers in a sequence: $(\forall \epsilon)(\exists \delta)(\forall x)$. To be precise, it is not just that there are three quantifiers in a sequence that makes this a hard sentence to understand, but the fact that the types of quantifiers used alternate.

Example 3.1.30. In this example, we describe how to interpret two quantifiers in sequence. Consider the following 8 sentences involving the 2-place predicate symbol P:

1. $(\forall x)(\forall y)P(x, y)$. 2. $(\forall x)(\exists y)P(x, y)$.
3. $(\exists x)(\forall y)P(x, y)$. 4. $(\exists x)(\exists y)P(x, y)$.
5. $(\forall y)(\forall x)P(x, y)$. 6. $(\exists y)(\forall x)P(x, y)$.
7. $(\forall y)(\exists x)P(x, y)$. 8. $(\exists y)(\exists x)P(x, y)$.

The first thing to note is that if all the quantifiers are of the same type then their order is immaterial. Thus (1)≡(5) and (4)≡(8). This will be proved when we study truth trees. We can therefore omit (5) and (8). We shall describe the meanings of the remaining sentences in two different interpretations below, but first let me make some general comments. The important point to remember at the outset when confronted with a sequence of quantifiers is always to work from left to right. If all the quantifiers are of the same type, as they are in (1) and (4), then they are easy to interpret; for (1) to be true in an interpretation $P(x, y)$ must be true whatever x and y are whereas for (2) to be true in an interpretation $P(x, y)$ must be true for at least one choice of x and y. The fun only starts when the quantifiers are not all of the same type, such as in (2), (3), (6) and (7). The sentence (2) means, in terms of directed graphs, that each vertex is at the base of an arrow, whereas (3) means that there is a vertex which has an arrow to every vertex (not excluding itself). The sentences (6) and (7) have similar meanings except that the arrows are coming in rather than going out. Now we consider two interpretations. In the first interpretation, the domain is $\mathbb{N}$, the set of natural numbers, and P is interpreted as $\leq$.

- $(\forall x)(\forall y)(x \leq y)$. This says that given any two natural numbers m and n we always have that $m \leq n$. This is false if $m = 2$ and $n = 1$.

- $(\forall x)(\exists y)(x \leq y)$. This says that for each natural number m there is a natural number n such that $m \leq n$ or, to paraphrase this in a more meaningful way, there is no largest natural number.

- $(\exists x)(\forall y)(x \leq y)$. This says there is a smallest natural number. This is true since 0 is that smallest natural number.

- $(\exists x)(\exists y)(x \leq y)$. This says that there are natural numbers m and n such that $m \leq n$. This is true since $1 \leq 1$.

- $(\exists y)(\forall x)(x \leq y)$. This says there is a largest natural number. This is false.

- $(\forall y)(\exists x)(x \leq y)$. This says that given any natural number n you can find a natural number m such that $m \leq n$. This is true since $n \leq n$.

In the second interpretation, the domain is 'all people' and P is interpreted by the binary relation 'is the father of'.

- $(\forall x)(\forall y)(x$ is the father of $y)$. This says that given any two people a and b then a is the father of b. This is false.

- $(\forall x)(\exists y)(x$ is the father of $y)$. This says that everyone is a father. This is false.

- $(\exists x)(\forall y)(x$ is the father of $y)$. This says that there is someone who is the father of everyone. This is false.

- $(\exists x)(\exists y)(x$ is the father of $y)$. This says that there are people a and b such that a is the father of b. This is true.

- $(\exists y)(\forall x)(x$ is the father of $y)$. This says that there is someone for whom everyone is a father. This is false.

- $(\forall y)(\exists x)(x$ is the father of $y)$. This says that everyone has a father. This is true.

Our final example is a famous one in the development of the theory of quantifiers since it shows how quantifiers can be used to disambiguate some English sentences. It is, really, part of Example 3.1.30 but is worth presenting on its own.

Example 3.1.31. Consider the sentence 'everyone loves someone'. If you reflect for a moment, you will discover that it is ambiguous. It could mean either (1) there is one person whom everyone loves or (2) each person loves someone or other (but not necessarily the same one). FOL enables us to separate these two meanings. Let L be a 2-place predicate symbol that we shall interpret as 'loves' where the domain is all people. Thus, $L(x, y)$ will be interpreted as 'x loves y'. Consider first the sentence $(\exists y)(\forall x)L(x, y)$. This is interpreted as saying 'there is one person whom everyone loves'. Consider next the sentence $(\forall x)(\exists y)L(x, y)$. This is interpreted as saying 'for each person there is someone whom they love'.

Exercises 3.1

1. In this question, use the following interpretation.[15]

   a: Athelstan.

   e: Ethelgiva.

   c: Cenric.

   $M(x)$: x is melancholy.

   $C(x)$: x is a cat.

   $L(x, y)$: x likes y.

   $T(x, y)$: x is taller than y.

   Transcribe the following formulae into colloquial English.

   (a) $\neg L(a, a)$.

   (b) $L(a, a) \rightarrow \neg T(a, a)$.

   (c) $\neg(M(c) \vee L(c, e))$.

   (d) $C(a) \leftrightarrow (M(a) \vee L(a, e))$.

   (e) $(\exists x)T(x, c)$.

   (f) $(\forall x)L(a, x) \wedge (\forall x)L(c, x)$.

   (g) $(\forall x)(L(a, x) \wedge L(c, x))$.

   (h) $(\exists x)T(x, a) \vee (\exists x)T(x, c)$.

---

[15]This question and the next were adapted from Chapter 1 of Volume II of [73].

(i) $(\exists x)(T(x,a) \vee T(x,c))$.

(j) $(\forall x)(C(x) \rightarrow L(x,e))$.

(k) $(\exists x)(C(x) \wedge \neg L(e,x))$.

(l) $\neg(\forall x)(C(x) \rightarrow L(e,x))$.

(m) $(\forall x)[C(x) \rightarrow (L(c,x) \vee L(e,x))]$.

(n) $(\exists x)[C(x) \wedge (M(x) \wedge T(x,c))]$.

2. Transcribe the following English sentences into formulae using the interpretation of Question 1.

(a) Everyone likes Ethelgiva.

(b) Everyone is liked by either Cenric or Athelstan.

(c) Either everyone is liked by Athelstan or everyone is liked by Cenric.

(d) Someone is taller than both Athelstan and Cenric.

(e) Someone is taller than Athelstan and someone is taller than Cenric.

(f) Ethelgiva likes all cats.

(g) All cats like Ethelgiva.

(h) Ethelgiva likes some cats.

(i) Ethelgiva likes no cats.

(j) Anyone who likes Ethelgiva is not a cat.

(k) No one who likes Ethelgiva is a cat.

(l) Somebody who likes Athelstan likes Cenric.

(m) No one likes both Athelstan and Cenric.

3. For each of the following formulae, draw its parse tree and show that it is closed and therefore a sentence.

(a) $(\forall x)P(x) \rightarrow (\exists x)P(x)$.

(b) $(\exists x)P(x) \rightarrow (\exists y)P(y)$.

(c) $(\forall y)((\forall x)P(x) \rightarrow P(y))$.

(d) $(\exists y)(P(y) \rightarrow (\forall x)P(x))$.

(e) $\neg(\exists y)P(y) \rightarrow [(\forall y)((\exists x)P(x) \rightarrow P(y))]$.

4. In the language of Example 3.1.4, write down formulae that correspond to the following relations in the interpretation.

(a) x is the sister of y.

(b) x is a grandfather.

(c) x is married.

5. A language consists of two 1-place predicates A and B and one 2-place predicate P. Describe models of the sentence

$$(\forall x)(A(x) \oplus B(x)).$$

3.2 Gödel's completeness theorem

In this section, we shall prove the main theorem of this book that will connect 'proof with truth' but before we can do this we need to generalize truth trees from PL to FOL and this turns out not to be straightforward.

3.2.1 An example

Our goal is to generalize Section 1.11 to FOL. So, we would like to develop a method that would tell us in a finite number of steps whether a sentence was satisfiable, that is, whether it had a model or not. This fails because of the following example.

Example 3.2.1. There are sentences that have no finite models. Put $S = F_1 \wedge F_2 \wedge F_3$ where

- $F_1 = (\forall x)(\exists y)R(x, y)$.

- $F_2 = (\forall x)\neg R(x, x)$.

- $F_3 = (\forall x)(\forall y)(\forall z)[R(x, y) \wedge R(y, z) \rightarrow R(x, z)]$.

Observe first that S does have a model. Define a structure to have as domain $\mathbb{N}$ equipped with the binary relation $<$. We therefore interpret $R(x, y)$ as $x < y$. Sentence F_1 is true in this interpretation since $m < m + 1$ for all $m \in \mathbb{N}$; sentence F_2 is true in this interpretation since $m \not< m$ for all $m \in \mathbb{N}$; sentence F_3 is true since if $m < n$ and $n < p$ then $m < p$. It follows that S is true in this interpretation. We now prove that S has *no finite models*; that is, there are no models in which the domain is finite. Let (D, ρ) be a model of S. Thus D is a non-empty set and ρ is a binary relation on D. Let $a_1 \in D$. Such an element exists because D is non-empty. Then, since F_1 is true it follows that there exists $a_2 \in D$ such that $a_1 \rho a_2$. Because F_2 holds it follows that $a_1 \neq a_2$. Thus D has at least 2 elements. We now start again. Since F_1 is true there exists $a_3 \in D$ such that $a_2 \rho a_3$. By F_2, we know that $a_2 \neq a_3$. But, from $a_1 \rho a_2$ and $a_2 \rho a_3$ we deduce by F_3 that $a_1 \rho a_3$. It follows by F_2 that $a_1 \neq a_3$. Thus the elements a_1, a_2, a_3 are distinct and so D has at least 3 elements. This process can be continued and leads to the conclusion that D must contain infinitely many elements and so S has no finite models.

Because of the above example, we shall revise what we would like truth trees for FOL to do.

In FOL, truth trees are used to show that a sentence has no model.

If we prove that Y has no model then we have, as a result, proved that $\neg Y$ is universally valid. We know that deciding universal validity is important since it lies behind showing that arguments in FOL are valid. Thus, given a sentence X, if the truth tree for $\neg X$ closes, then X is universally valid. Ideally, we would like this to be an algorithm for deciding whether X has no models or not, but this will turn out to be impossible, as we shall see.[16]

3.2.2 Truth trees for FOL

The goal of truth trees for FOL is to show that the formula placed at the root of the tree has no models, and we would hope to show this by the tree closing in a finite number of steps. The motivating idea behind the truth tree procedure is that we attempt to construct a model for the formula but fail because of the presence of what we might call *internal contradictions*. The question is: how do we construct such a model? The answer is: from the very language itself augmented by any extra constants that might be needed. The constants will form the elements of the structure, and will do double duty as constants, whilst the predicate symbols will do double duty and stand for their interpreted predicates. This might sound paradoxical, a bit like Baron Munchausen pulling himself up by his own hair, but it does all work. I will begin by listing the rules of truth trees for FOL. It is helpful to think of the following rules in the following terms: *convert FOL sentences into PL wff by means of substitution by constants.*

---

[16]Since, within the murky depths of FOL, infinity lurks.

Truth tree rules for FOL

- All PL truth tree rules are carried forward.

- De Morgan's rules for quantifiers:

$$\neg(\forall x)X\checkmark \qquad\qquad \neg(\exists x)X\checkmark$$
$$(\exists x)\neg X \qquad\qquad (\forall x)\neg X$$

- New-name rule:

$$(1)\ (\exists x)X[x]\checkmark$$
$$X[a]$$

where we add $X[a]$ at the bottom of all branches containing (1) and a is a constant that does not already appear in the branch containing (1).

- Never-ending rule:

$$(2)\ (\forall x)X[x]\ *$$
$$X[a]$$

where we add $X[a]$ at the bottom of a branch containing (2) and a is any constant appearing in the branch containing (2) or a is a new constant if no constants have yet been introduced. We have used the $*$ to mean that the formula is never used up.

We can use truth trees for FOL in the same way as in PL:

- To prove that $\vDash X$ we show that some truth tree with root $\neg X$ closes.

- To prove that $X_1,\ldots,X_n \vDash X$ we show that some truth tree with root $X_1,\ldots,X_n,\neg X$ closes.

- To prove that $X \equiv Y$ we show that $\vDash X \leftrightarrow Y$.

Only the new-name rule and the never-ending rule need any comment since the other rules are straightforward.

The new-name rule

Recall that given a formula X, we are trying to construct a model of X but that model will be built out of the language itself. Let $X = (\exists x)Y$, where x is free in Y. Then we give a unique name to the element that is supposed to make Y true when all free occurrences of x in Y are replaced by that name. The name we choose must be a previously unused constant whose sole function is to name the element that is supposed to satisfy $(\exists x)Y$.

Example 3.2.2. We do this all the time in mathematics. For example, we give the name $\sqrt{2}$ to the real number that is supposed to make the sentence $X = (\exists x)[(x \geq 0) \wedge (x^2 = 2)]$ true when interpreted in the real numbers. Thus we know that $\sqrt{2} \geq 0$ and that $(\sqrt{2})^2 = 2$ but that is *all* we know about the symbol $\sqrt{2}$.

We may therefore replace $(\exists x)Y$ by $(x \leftarrow a)Y$, where a is the new constant. Remember that the only thing we know about a is that it is supposed to make $(x \leftarrow a)Y$ true. The formula $(x \leftarrow a)Y$ will gradually be decomposed by the truth tree rules and as this happens we will obtain more explicit information about the properties that a is supposed to have. It is in this way that internal contradictions can come to light. But why a new name? Because if we used a previous name it would come with baggage, laden with other properties that had nothing to do with its simply satisfying $(\exists x)Y$.

The never-ending rule

Let $X = (\forall x)Y$, where x is free in Y. Now, X is supposed to be true for *every* element of the domain of the interpretation. Consequently, every time a new constant a appears we must replace all free occurrences of x in Y by a because whatever X is asserting it is certainly asserting it about a. Thus, the never-ending rule is never used up, which is why I have marked it in the above algorithm with a $*$ rather than a $\checkmark$. The sentence at the root of the tree could contain constants already and these would also have to satisfy X. If there are no such constants in the language then we can get the ball rolling by producing a starting constant a_1 out of nothing and then writing down $(x \leftarrow a_1)Y$. We can do this because our domains are always non-empty.

> A good strategy in constructing a truth tree using the above rules is apply the new-name rule first before applying the never-ending rule, whenever this is possible.

There are a couple of other issues that arise when constructing truth trees in FOL that do not arise in PL.

Examples 3.2.3.

1. Truth trees can have infinite branches. Here is an example that arises from applying the never-ending rule and the new-name rules in sequence; I have numbered the lines so that I can refer to them below.

$$(1)\ (\forall x)(\exists y)R(x, y)\ *$$
$$|$$
$$(2)\ (\exists y)R(a_1, y)\ \checkmark$$
$$|$$
$$(3)\ R(a_1, a_2)$$
$$|$$
$$(4)\ (\exists y)R(a_2, y)\ \checkmark$$
$$|$$
$$(5)\ R(a_2, a_3)$$
$$|$$

and so on

Observe that in line (1) we get the ball rolling by applying the never-ending rule with the constant a_1 that is produced out of thin air to obtain line (2). We then invoke the new-name rule with the constant a_2 to obtain line (3). But then we must invoke the never-ending rule again in line (1) with respect to the new constant a_2 since, whatever $(\forall x)(\exists y)R(x, y)$ is saying, it must be saying it about a_2. This gives us line (4). But now we must invoke the new-name rule with the new constant a_3 to obtain line (5). We are therefore trapped in an endless loop that forces us to generate an infinite sequence of constants $a_1, a_2, a_3, \ldots$ as new names.

2. The order in which the rules for truth trees are applied in FOL does matter. If we place the formulae $(\forall x)(\exists y)P(x, y)$ and $P(a) \wedge \neg P(a)$ at the root, then we can get an infinite tree if we repeatedly apply the tree rules to the first formula, but an immediate contradiction, and so a closed finite truth tree, if we start with the second formula instead.

The truth tree method described in this section is not an algorithm but a tool. For example, the order in which you apply the rules matters. This means that you have to look ahead and use insight to attempt to close a tree. One of the ways of doing this is to make choices that are likely to cause contradictions.

Examples 3.2.4.

1. We show that the following is a valid argument:

$$(\forall x)(H(x) \to M(x)), H(a) \vDash M(a).$$

This, of course, is just the abstract form of the argument that began this chapter. The language that we are using here has two 1-place predicate symbols and one constant a. Here is the truth tree:

$$(\forall x)(H(x) \to M(x)) *$$
$$H(a)$$
$$\neg M(a)$$
$$|$$
$$H(a) \to M(a) \checkmark$$

$$\diagup \diagdown$$
$$\text{✗} \neg H(a) \quad M(a) \text{ ✗}$$

Observe that because the constant a is available in the language, we can use it in our first application of the never-ending rule. The truth tree closes and so the original argument is valid.

2. We show that the following argument is valid:

$$(\exists x)(\forall y)F(x, y) \vDash (\forall y)(\exists x)F(x, y).$$

Here is the truth tree:

$$(1)\ (\exists x)(\forall y)F(x, y)\ \checkmark$$
$$(2)\ \neg(\forall y)(\exists x)F(x, y)\checkmark$$
$$|$$
$$(3)\ (\exists y)\neg(\exists x)F(x, y)\checkmark$$
$$|$$
$$(4)\ (\exists y)(\forall x)\neg F(x, y)\checkmark$$
$$|$$
$$(5)\ (\forall y)F(a, y) *$$
$$|$$
$$(6)\ (\forall x)\neg F(x, b) *$$
$$|$$
$$(7)\ F(a, b)$$
$$|$$
$$(8)\ \neg F(a, b)\ \text{✗}$$

I have numbered the lines so I can refer to them below. Line (3) is obtained from line (2) by pushing the negation through the quantifier. Line (4) is obtained from line (3) by pushing the negation through the quantifier. I would usually just write down line (4) immediately from line (2) to save time. We now apply the new-name rule to line (1), with the new constant a to obtain line (5), and to line (4), with the new constant b to obtain line (6). Having used up any possible applications of the new-name rule, we can now apply the never-ending rule. This is done first to line (5) to obtain line (7), and then to line (6) to obtain line (8). The choice of constants I used in each case was dictated by the fact that I was trying to obtain a contradiction. This is exactly what I have achieved in lines (7) and (8). The truth tree therefore closes and so the argument is valid.

You might wonder why we don't also try to prove

$$(\forall y)(\exists x)F(x,y) \vDash (\exists x)(\forall y)F(x,y).$$

The reason is that it isn't true. To see why, we construct a model of $(\forall y)(\exists x)F(x,y)$ in which $(\exists x)(\forall y)F(x,y)$ is false. In other words: we construct a counterexample. Since F is a 2-place predicate symbol, the structure that does the job can be represented by means of a directed graph as in Section 3.1.2. Observe that the structure below is a model of $(\forall y)(\exists x)F(x,y)$ since each vertex is at the pointed end of an arrow. It is not, however, a model of $(\exists x)(\forall y)F(x,y)$ because this would require a directed graph in which there was a vertex that pointed to every vertex (including itself).

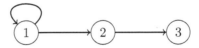

3. We prove that

$$\vDash (\exists x)(A(x) \wedge B(x)) \rightarrow ((\exists x)A(x) \wedge (\exists x)B(x)).$$

Here is the truth tree:

$$\neg((\exists x)(A(x) \wedge B(x)) \rightarrow ((\exists x)A(x) \wedge (\exists x)B(x))) \checkmark$$
$$|$$
$$(\exists x)(A(x) \wedge B(x)) \checkmark$$
$$\neg((\exists x)A(x) \wedge (\exists x)B(x)) \checkmark$$
$$|$$
$$A(a) \wedge B(a) \checkmark$$
$$|$$
$$A(a)$$
$$B(a)$$

$$\neg(\exists x)A(x) \qquad \neg(\exists x)B(x)$$
$$|$$
$$(\forall x)\neg A(x) * \qquad (\forall x)\neg B(x) *$$
$$|$$
$$\neg A(a)\ \textbf{✗} \qquad \neg B(a)\ \textbf{✗}$$

The truth tree closes and so the sentence is universally valid.

It is not true, however, that

$$((\exists x)A(x) \wedge (\exists x)B(x)) \rightarrow (\exists x)(A(x) \wedge B(x))$$

is universally valid. We construct a counterexample. To do this, we construct a model of $((\exists x)A(x) \wedge (\exists x)B(x))$ in which $(\exists x)(A(x) \wedge B(x))$ is false. Since A and B are 1-place predicate symbols, the structure that does the job should be a set equipped with two subsets. Define $D = \{1, 2\}$ and $X = \{1\}$ and $Y = \{2\}$. Interpret A by X and B by Y. Then the sentence $(\exists x)A(x) \wedge (\exists x)B(x)$ is true in this interpretation because it simply says that X and Y are non-empty. The sentence $(\exists x)(A(x) \wedge B(x))$ is false in this interpretation because it says that X and Y have a non-empty intersection, but in fact X and Y are disjoint.

4. We prove that

$$(\forall x)(\forall y)A(x, y) \equiv (\forall y)(\forall x)A(x, y).$$

Here is the truth tree:

$$\neg((\forall x)(\forall y)A(x, y) \leftrightarrow (\forall y)(\forall x)A(x, y)) \checkmark$$

$$(\forall x)(\forall y)A(x, y) * \qquad \neg(\forall x)(\forall y)A(x, y)$$
$$\neg(\forall y)(\forall x)A(x, y) \checkmark \qquad (\forall y)(\forall x)A(x, y) *$$
$$\mid \qquad\qquad\qquad \mid$$
$$(\exists y)(\exists x)\neg A(x, y) \checkmark \qquad \text{and similarly}$$
$$\mid$$
$$(\exists x)\neg A(x, a) \checkmark$$
$$\mid$$
$$\neg A(b, a)$$
$$\mid$$
$$(\forall y)A(b, y) *$$
$$\mid$$
$$A(b, a) \; \boldsymbol{\mathsf{X}}$$

The truth tree closes and so the two sentences are, indeed, logically equivalent. This result can be generalized to one which says that the order in which a sequence of universal quantifiers is written doesn't matter and, similarly, the order in which a sequence of existential quantifiers is written doesn't matter. As we have seen, though, if existential and universal quantifiers are mixed the order in which they occur really does matter.

3.2.3 The soundness theorem

We have assumed throughout the previous section that if a truth tree having $\neg X$ as a root closes, then X is universally valid. This now needs to be proved but to do so requires some preparation. Let me recall some terminology.

An α-*sentence* X is a sentence that has one of the following forms: $X_1 \wedge X_2$; $\neg(U \vee V) \equiv X_1 \wedge X_2$, where $X_1 = \neg U$ and $X_2 = \neg V$; $\neg(U \rightarrow V) \equiv X_1 \wedge X_2$, where $X_1 = U$ and $X_2 = \neg V$; $\neg\neg U \equiv X_1$, where $X_1 = X_2 = U$. A β-*sentence* is a sentence that has one of the following forms: $X_1 \vee X_2$; $\neg(U \wedge V) \equiv X_1 \vee X_2$, where $X_1 = \neg U$ and $X_2 = \neg V$; $U \rightarrow V \equiv X_1 \vee X_2$, where $X_1 = \neg U$ and $X_2 = V$; $U \leftrightarrow V \equiv X_1 \vee X_2$, where $X_1 = U \wedge V$ and $X_2 = \neg U \wedge \neg V$; $\neg(U \leftrightarrow V) \equiv X_1 \vee X_2$, where $X_1 = U \wedge \neg V$ and $X_2 = \neg U \wedge V$. Thus in the context of truth trees, an α-sentence is a sentence to which an α-rule may be applied and a β-sentence is a sentence to which a β-rule may be applied.

The key step in the proof of the soundness theorem uses the idea of the expansion of a language L by adjoining constants described in Section 3.1.5. If L is expanded to the language L' by adjoining constants in the set U, say, then any interpretation I of L can be expanded to an interpretation I' of L' by interpreting the elements of U as additional distinguished elements.

The following is similar to Lemma 1.11.15 for PL.

Lemma 3.2.5. *Let $\mathscr{B}$ be a set of sentences in the language L which has D as a model.*

1. *Let X be an α-sentence in $\mathscr{B}$. Then D is a model of $\mathscr{B} \cup \{X_1, X_2\}$.*

2. *Let X be a β-sentence in $\mathscr{B}$. Then D is a model of either $\mathscr{B} \cup \{X_1\}$ or $\mathscr{B} \cup \{X_2\}$.*

3. *Let X be a sentence in $\mathscr{B}$ of the form $(\exists x)Y[x]$. Then there is a 1-step expansion of the language L to a language L' by adjoining a new constant a and there is a 1-step expansion of the model D which is an interpretation of the language L' and a model of $\mathscr{B} \cup \{(x \leftarrow a)Y[x]\}$.*

4. *Let X be a sentence in $\mathscr{B}$ of the form $(\forall x)Y[x]$. There are two cases:*

   (a) *The language L contains no constants. Then there is a 1-step expansion of the language L to a language L' by adjoining a new constant a and there is a 1-step expansion of the model D which is an interpretation of the language L' and a model of $\mathscr{B} \cup \{(x \leftarrow a)Y[x]\}$.*

   (b) *The language L contains at least one constant. Then D is also a model of $\mathscr{B} \cup \{(x \leftarrow a)Y[x] :$ for all constants a in $L\}$.*

Proof. (1) The structure D is a model of X and so it is also a model of both X_1 and X_2.

(2) The structure D is a model of X and so it is also a model of at least one of X_1 or X_2.

(3) The structure D is a model of $X = (\exists x)Y[x]$. Let the subset of D determined by $Y[x]$ be denoted by $\mathcal{Y}$. Then there is an element $d \in D$ such that $d \in \mathcal{Y}$. Add the element d to the structure D as a distinguished element to obtain the structure D' and interpret a as d. Then the structure D' is an interpretation of the language L' in which all the formulae in $\mathscr{B} \cup \{(x \leftarrow a)Y[x]\}$ are true.

(4) Case (a). Suppose first that L has no constants. Let the subset of D determined by $Y[x]$ be denoted by $\mathcal{Y}$. Since the domain is non-empty, there is certainly an element $d \in D$ such that $d \in \mathcal{Y}$ is true. Add the element d to the structure D as a distinguished element to obtain the structure D' and interpret a as d. Then the sentence $Y[a]$ has D' as a model. Thus a 1-step expansion of D is a model of both $\mathcal{B}$ and $(x \leftarrow a)Y[x]$.

Case (b). Suppose that L already has constants. Then $Y[a]$ will be true in the given model for any constant a in the language. Thus D is a model of all the formulae in $\mathcal{B}$ and all the formulae $(x \leftarrow a)Y[x]$ where a is a constant in L. $\qquad\square$

Theorem 3.2.6 (Soundness theorem). *Let X be a sentence. If some truth tree for $\neg X$ closes then X is universally valid.*

Proof. Suppose that some truth tree $\mathcal{T}$ for $\neg X$ closes but X is not universally valid. Then $\neg X$ has a model D. By repeatedly applying Lemma 3.2.5, it follows that $\mathcal{T}$ must have a branch such that all formulae occurring on that branch are true in some expansion of D. But such a branch cannot be closed which contradicts our assumption. It follows that X is universally valid. $\qquad\square$

3.2.4   The completeness theorem

Up to now, ingenuity has been required to construct truth trees for FOL. The obvious question, therefore, is whether the method of truth trees can be turned into a proper algorithm for deciding whether a sentence is universally valid or not. This question is known by its German name as the *Entscheidungsproblem*. The answer to this question is not only 'no' but, in fact, a 'resounding no' since there is no algorithm, of any complexion, for deciding whether a sentence is universally valid or not. This is therefore quite different from the analogous question of asking whether a wff in PL is a tautology or not. This was proved independently in 1936 by Alonzo Church in the US and by Alan Turing in the UK. Turing's resolution to this question was far-reaching since it involved him in formulating, mathematically, what we would now call a computer.

It is, however, possible to improve upon the truth tree method described in the previous section by converting it into what is called the *systematic* truth tree method. Such trees lead to a result know as *Gödel's completeness theorem*, proved by Kurt Gödel (1906–1978),[17] which is the basis for all further work in FOL. However, the drawback of systematic truth trees is that they are badly adapted to manual calculations so we only use them in this section to prove the big theorem.

We need a measure of the complexity of a formula X in FOL just as we did for wff in PL. The most obvious way to do this is to simply count the

---

[17] Often described as the greatest logician since Aristotle.

number of propositional connectives *and* quantifiers in X. This is called the *degree* of X. Thus the formulae of degree 0 are precisely the atomic formulae.

We now generalize the Hintikka sets encountered in Section 1.11. Let $\mathscr{H}$ be a set of sentences using the set $\mathscr{P}$ of constants. We say that $\mathscr{H}$ is a *Hintikka set* if it satisfies the following properties:

(H1) No atomic formula and its negation both belong to $\mathscr{H}$.

(H2) For any α-sentence X in $\mathscr{H}$ both X_1 and X_2 belong to $\mathscr{H}$.

(H3) For any β-sentence X in $\mathscr{H}$ either X_1 or X_2 belong to $\mathscr{H}$.

(H4) For every sentence in $\mathscr{H}$ of the form $(\exists x)X[x]$ then $X[a]$ is in $\mathscr{H}$ for some constant $a \in \mathscr{P}$.

(H5) For every sentence in $\mathscr{H}$ of the form $(\forall x)X[x]$ then $X[a]$ is in $\mathscr{H}$ for all constant $a \in \mathscr{P}$.

The key property of Hintikka sets is the following which tells us how to construct a model from purely syntactic information. The proof is similar to that of Lemma 1.11.13.

Proposition 3.2.7. *Let $\mathscr{H}$ be a Hintikka set. Then there is an interpretation in which all elements of $\mathscr{H}$ are true.*

Proof. Define the domain of the structure, D, to be the set of all constants $\mathscr{P}$. Each constant $a \in \mathscr{P}$ is interpreted as itself. For each n-ary predicate symbol P that occurs in some formulae in $\mathscr{H}$, define ρ_P to be an n-ary relation defined on D where $(a_{i_1}, \ldots, a_{i_n}) \in \rho_P$ if and only if $P(a_1, \ldots, a_n) \in \mathscr{H}$. Observe that by (H1), our definition of ρ_P makes sense. We have therefore defined a structure which we call the *natural structure* associated with our Hintikka set. The claim is that all the sentences in $\mathscr{H}$ are true when interpreted in the natural structure. Atomic formulae have degree 0 and they are all true by construction.

Suppose that we have proved that all sentences in $\mathscr{H}$ of degree at most n are true in the natural structure. Let $X \in \mathscr{H}$ be a sentence of degree $n + 1$. We shall prove that it, too, is true in the natural structure. If X is an α- or β-sentence then this is straightforward so we just need to deal with sentences beginning with quantifiers. Suppose first that $X = (\forall x)Y[x]$. Then by the definition of a Hintikka set, $Y[a] \in \mathscr{H}$ for all $a \in \mathscr{P}$ and all the sentences $Y[a]$ are true in the natural structure since they have degree n. But this implies that $X = (\forall x)Y[x]$ is true in the natural structure. Suppose now that $X = (\exists x)Y[x]$. Then by the definition of a Hintikka set, $Y[a] \in \mathscr{H}$ for some $a \in \mathscr{P}$ and the sentence $Y[a]$ is true in the natural structure since it has degree n. But this implies that $X = (\exists x)Y[x]$ is true in the natural structure. $\square$

Example 3.2.8. We look again at part (1) of Examples 3.2.3. Let $\mathscr{H}$ be the set of formulae that occur as vertices of this infinite branch. Thus

$$\mathscr{H} = \{(\forall x)(\exists y)R(x, y), (\exists y)R(a_i, y), R(a_i, a_{i+1}) \colon i \geq 1\}.$$

184

It is easy to check that $\mathscr{H}$ is a Hintikka set. The model described by Proposition 3.2.7 is then simply

$$\boxed{a_1} \longrightarrow \boxed{a_2} \longrightarrow \boxed{a_3} \longrightarrow \boxed{a_4} \qquad \cdots$$

Example 3.2.9. Hintikka sets can also be finite. Put

$$X = \neg[(\forall x)(P(x) \vee Q(x)) \rightarrow ((\forall x)P(x) \vee (\forall x)Q(x))].$$

Let

$$
\begin{aligned}
\mathscr{H} \;=\; & \{X, (\forall x)(P(x) \vee Q(x)), \neg[(\forall x)P(x) \vee (\forall x)Q(x)], \neg(\forall x)P(x), \\
& \neg(\forall x)Q(x), (\exists x)\neg P(x), (\exists x)\neg Q(x), \neg P(a_1), \neg Q(a_2), \\
& P(a_1) \vee Q(a_1), Q(a_1), P(a_2) \vee Q(a_2), P(a_2)\}.
\end{aligned}
$$

You can check that this is a Hintikka set. Put $D = \{a_1, a_2\}$. The natural structure associated with $\mathscr{H}$ is

$$(\{a_1, a_2\}, \{a_1\}, \{a_2\})$$

where P is interpreted as the subset $\{a_2\}$ and Q is interpreted as the subset $\{a_1\}$. We now check that the natural structure is a model of the sentence X. Observe first that each element of D belongs either to the subset $\{a_1\}$ or the subset $\{a_2\}$. It follows that the natural structure is a model of the sentence $(\forall x)(P(x) \vee Q(x))$. However, the natural structure is not a model of the sentence $(\forall x)P(x) \vee (\forall x)Q(x)$ because neither of the two subsets $\{a_1\}$ or $\{a_2\}$ is equal to D. This proves that the natural structure is a model of X.

Before we state the next result, we give an example of what, from our perspective, is bad behaviour.

Example 3.2.10. Here is a tree with infinitely many vertices but no infinite branch:

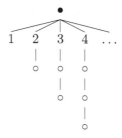

Observe that the root vertex of the above tree, marked in black, has infinitely many successors labelled $1, 2, 3, 4, \ldots$. Against this, a tree in which each vertex has a finite number of successors is said to be *finitely generated*.

Lemma 3.2.11 (König's lemma). *A finitely generated tree with infinitely many vertices has an infinite branch.*

Proof. Call a vertex *good* if it has infinitely many descendants and *bad* if it does not. The root is good. Now observe that each good vertex must have at least one good successor because it has only finitely many successors. Choose the leftmost one. Repeat this procedure. In this way, we find an infinite branch. $\square$

We now describe *systematic truth trees* by making two amendments to the truth tree method described in the previous section:

1. *We construct the truth tree in a more organized way.*

    - A check mark is placed next to a formula to show that it has been *used*.

    - Each stage of the systematic truth tree algorithm will begin by finding the *first unused formula on an open branch*; this is located by starting at the root and working down the tree from left-to-right; we therefore carry out a breadth-first search.

    - The constants $a_1, a_2, a_3, \ldots$ should be regarded as being ordered so that we can always refer to the first *unused* constant.

2. *We modify the never-ending rule.* For *every* open branch $\mathscr{B}$ containing an unused occurrence of $(\forall x)X[x]$ at vertex v, do the following: if a is the first (remember the ordering!) constant such that $X[a]$ is not in $\mathscr{B}$ append $(\forall x)X[x], X[a]$ to $\mathscr{B}$ as a leaf; show that the occurrence of $(\forall x)X[x]$ at vertex v is used by putting a check mark against it.

The systematic truth tree algorithm now operates just like its unsystematic sibling except that at each stage of the algorithm we locate the first unused formula on an open branch. We say that such a truth tree is *finished* if it is infinite or cannot be extended further by applying any of the systematic truth tree rules. The rationale for the introduction of systematic truth trees is provided by the following.

Lemma 3.2.12. *In a finished systematic truth tree, each open branch is a Hintikka set.*

Proof. Let $\mathscr{B}$ be an open branch. Since it is open, we know that (H1) holds. The important point to observe next is that because of the notion of 'used formula' and the breadth-first search procedure that we are using, each formula on the tree will be dealt with eventually — this obviates the problem of formulae being ignored such as in the second of the two examples in Examples 3.2.3. This means that (H2)–(H4) all hold. To show that (H5) holds, simply observe that every time the never-ending rule is applied to a universally quantified formula $(\forall x)X[x]$ in a branch, a new constant a appears at the leaf of that branch as $(x \leftarrow a)X[x]$, a copy of $(\forall x)X[x]$ is made at that leaf and the previous occurrence of $(\forall x)X[x]$ is marked as used. We therefore copy the formula $(\forall x)X[x]$ at the leaves of all open branches that contain it. In this way, it is never switched off and leads to each constant in turn being introduced in every open branch. $\square$

We can now state and prove the companion to the soundness theorem.

Theorem 3.2.13 (Completeness theorem). *Let X be a universally valid sentence. Then a systematic truth tree having $\neg X$ as root must close.*

Proof. Suppose that a finished systematic truth tree having $\neg X$ as root doesn't close. Then it is either finite, and has an open branch, or infinite, and by Lemma 3.2.11, has an infinite open branch. In either case, the open branch is a Hintikka set by Lemma 3.2.12 and so has a model by Proposition 3.2.7. Thus $\neg X$ has a model contradicting the fact that X is universally valid. □

We introduce some notation to state precisely what we have proved in this, and the previous, section. We write $\vdash X$ if some truth tree with root $\neg X$ closes. The symbol $\vdash$ is called *turnstile* and should be distinguished from the symbol $\vDash$ which, you will recall, is called *semantic turnstile*. If we combine Theorem 3.2.6 and Theorem 3.2.13, we obtain the following.

Theorem 3.2.14 (Gödel's completeness theorem). *Let X be a sentence. Then $\vdash X$ if and only if $\vDash X$.*

The symbol $\vdash$ is about proof and the symbol $\vDash$ is about truth. A proof of a formula X is a finite thing: it is a finite closed truth tree with $\neg X$ as its root. Given such a tree, you can quickly check that it has been constructed correctly.[18] To establish the truth of X we have, in principle, to search all interpretations of X and show that X is true in every one of them; this is an infinite process. What the above theorem therefore says is that proof — a finite thing — is a route to truth — which is an infinite thing.

You might think that the above theorem would actually prove the *Entscheidungsproblem*, but it does no such thing. To understand why, let X be a sentence and let's consider what happens when we construct the systematic truth tree having $\neg X$ as its root. If X is universally valid, then a systematic truth tree will close in a finite number of steps, and so all is well and good. If X is not universally valid, then there are two possibilities: either (1) the systematic truth tree can be constructed in a finite number of steps in which case there will be at least one open branch that will lead to a finite model for $\neg X$; or (2) the construction of the systematic truth tree never terminates. It is possibility (2) that scotches solving the *Entscheidungsproblem*. This may, of course, simply be an issue with systematic truth trees, but Turing proved that there is *no algorithm at all* that can solve the *Entscheidungsproblem*.

---

[18]Truth trees really are proofs although slightly disguised. This is the point of Section 1.12 which generalizes to the case of FOL.

Exercises 3.2

1. Prove that $(\forall x)R(x) \models (\exists x)R(x)$. Explain informally why it is a valid argument.

2. Prove that $(\forall x)(\forall y)R(x, y) \models (\forall x)R(x, x)$.

3. Prove that $(\exists x)(\exists y)F(x, y) \models (\exists y)(\exists x)F(x, y)$.

4. Prove that $(\forall x)(P(x) \wedge Q(x)) \equiv ((\forall x)P(x) \wedge (\forall x)Q(x))$.

5. Prove that $(\exists x)(P(x) \vee Q(x)) \equiv ((\exists x)P(x) \vee (\exists x)Q(x))$.

6. Prove that $((\forall x)P(x) \vee (\forall x)Q(x)) \to (\forall x)(P(x) \vee Q(x))$ is universally valid. On the other hand, show that the following is not universally valid by constructing a counterexample

$$(\forall x)(P(x) \vee Q(x)) \to ((\forall x)P(x) \vee (\forall x)Q(x)).$$

7. Prove that the formula $(\exists x)[D(x) \to (\forall y)D(y)]$ is universally valid. Interpret $D(x)$ as 'x drinks' with domain people. What does the above formula say in this interpretation? Does this seem plausible? Resolve the issue. *This is a famous example of Raymond Smullyan and the formula is known as 'Smullyan's drinking formula'.*

8. Let R be a 2-place predicate symbol. We say that an interpretation of R has the respective property (in italics) if it is a model of the corresponding sentence below:

   - *Reflexive:* $(\forall x)R(x, x)$.
   - *Irreflexive:* $(\forall x)\neg R(x, x)$.
   - *Symmetric:* $(\forall x)(\forall y)(R(x, y) \to R(y, x))$.
   - *Asymmetric:* $(\forall x)(\forall y)(R(x, y) \to \neg R(y, x))$.
   - *Transitive:* $(\forall x)(\forall y)(\forall z)(R(x, y) \wedge R(y, z) \to R(x, z))$.

   Illustrate these definitions by using directed graphs.

   Prove the following using truth trees.

   (a) If R is asymmetric then it is irreflexive.
   (b) If R is transitive and irreflexive then it is asymmetric.

3.3 Proof in mathematics (II)

> *If they give you ruled paper, write the other way.* — Juan Ramón
> Jiménez (quote by Ray Bradbury in *Fahrenheit 451*).

Natural deduction

If you open a book of mathematics, you will not, in general, see truth trees. It is not that logic is absent from mathematics, far from it, it is just that it is handled differently when it is *being used* from when it it is *being discussed*. Thus, when you see the word *Proof* in this book, what follows involves logic being used. However, to understand how logic is used in mathematics requires practice and, in my view, the best way to do this is to study mathematics and acquire an understanding of logic through the mathematics. It is certainly true that background study of the main ideas of logic, such as the meaning of the propositional connectives and the quantifiers, for example, is useful — this is one of the motivations of this book — but just as you cannot learn to swim from a book, so you cannot learn logic in the abstract. There is a formalization of first-order logic, called *natural deduction*, which arises from the way that we use logic in mathematics, but it is natural only if you are familiar with proofs in mathematics. Since I am not assuming that you are, I shall not describe this formalization here, but I will give a couple of examples to provide a taste of what it is like. If you like solving Sudoku, it will feel familiar. Both examples below are from [20].

Example 3.3.1. We prove that $X = (x \vee (y \wedge z)) \rightarrow ((x \vee y) \wedge (x \vee z))$ is a tautology in the way that would be most natural to a mathematician. If $x \vee (y \wedge z)$ were false then X would be true, irrespective of the truth value of $(x \vee y) \wedge (x \vee z)$. So, we can suppose that $x \vee (y \wedge z)$ is true. This means that either (1) x is true or (2) $y \wedge z$ is true. We deal with each of these two possibilities in turn. In case (1), the fact that x is true means that both $x \vee y$ and $x \vee z$ are true. Consequently, $(x \vee y) \wedge (x \vee z)$ is true. In case (2), the fact that $y \wedge z$ is true means that both y and z are true. This means that both $x \vee y$ and $x \vee z$ are true. Consequently, $(x \vee y) \wedge (x \vee z)$ is true. We have therefore proved that X is always true, and so X is a tautology.

Example 3.3.2. We prove that $Y = ((\exists x)(\forall y)F(x,y)) \rightarrow ((\forall y)(\exists x)F(x,y))$ is universally valid. With the same rationale as in Example 3.3.1, we shall assume that $(\exists x)(\forall y)F(x,y)$ is true and then prove that $(\forall y)(\exists x)F(x,y)$ is thereby true. By assumption, there is an x, a specific x, such that $F(x,y)$ is true for all y. We give a name to this specific x and call it a. It follows that $F(a,y)$ is true for all y. Using the meaning of the existential quantifier, it follows that $(\exists x)F(x,y)$ is true for all y. Finally, using the meaning of the universal quantifier, we deduce that $(\forall y)(\exists x)F(x,y)$ is true. We have therefore proved that Y is universally valid.

Natural deduction is well described (free, gratis and for nothing) in [73, Volume 1, Chapters 4, 5 and 7; Volume 2, Chapters 5 and 6].

Creativity

To conclude, it seems appropriate to return to the mathematics itself which served as the original motivation for the logic described in this book. Keep in mind the following analogy: between the act of creating music and the musical score that results. The process of creation and the process of, for want of a better word, notation, are not unrelated but they are clearly very different in character. Logic belongs to the notation side of mathematics more than it belongs to the creation side. I shall tell you, in a moment, *what* a proof is but I will not tell you how to *find* a proof. The process of finding a proof is one of creation for which there are no rules, as such, but rather guidelines and strategies. Once a proof has been constructed, however, then in principle anyone who is suitably trained can check it (and, perhaps one day, this will be something that will be automated).

Each field of mathematics is described by giving a set of *axioms* that together define the field. They serve as the 'rules of the game'.

> Constructing an axiom system to define an area of mathematics is part of the creative side of mathematics.

Once an axiomatic system has been set up the goal is to *prove* that certain statements follow from these axioms. A statement that does follow from the axioms is called a *theorem* of that axiomatic system.

> In reality, the goal is actually to prove *interesting* theorems but what that means and how this is achieved is again part of the creative side of mathematics.

The way that we prove a theorem is to write down a *proof* of that theorem. A proof is a sequence of statements, where the last statement is the one we are trying to prove, in which each statement is one of the following:

- an axiom,

- an assumption,

- a definition,

- a previously proved statement,

- a previously used statement,

- a statement that *follows from* some of the statements that precede it.

One of the goals of this book is to make sense of the phrase 'follows from' in the setting of first-order logic. The above notion of proof is exemplified by classical Euclidean geometry which begins with a handful of very simple axioms, such as 'two distinct points determine a line', which then lead to theorems such as that of Pythagoras which are interesting, useful and not at all simple.[19] Although a proof is written as a sequence, the way the proof was discovered will be anything but sequential. To return to the musical analogy I began this section with, a musical score is, at its most basic, a sequence of notes played in time, but there is no reason to suppose that the composer of the music actually dreamt it up in order from beginning to end. In addition, proofs of theorems often make radical departures to achieve their goals. A famous example is the proof of Fermat's conjecture which states that the equation $x^n + y^n = z^n$, where $n \geq 3$, has no integer solutions except when $xyz = 0$. Early attempts at proving this claim worked within number theory, which is the language the problem is couched in. The eventual proof converted it into a question in complex geometry.

I also need to stress the fact that the logic of this book is first-order but not all mathematics need be first-order.[20] For example, the statement 'every non-empty subset of the natural numbers contains a minimum element' is not a first-order statement since quantification is not over individual elements of the natural numbers but over subsets of the natural numbers; it is, therefore, a second-order statement. There is nothing wrong with second-order logic except that it cannot be formalized in the same way as first-order [7, Chapter 22].

3.4 Further reading

There are a number of classic, older works on logic that are still worth reading. First and foremost, there is Hilbert and Ackermann's pioneering book [30] but two others worth perusing are Kleene [40] and Quine [63].

Books on logic at roughly the same level as this one are Jeffrey [38], Mundici [57], Smullyan [69], Stoll [71], Teller [73] and Zegarelli [81].

Boolean algebras appear throughout mathematics. Although named after Boole, the relationship between Boole's work [6] and Boolean algebras is complex [25], but this is not surprising since new ideas take time before they assume their final form. A good place to start learning about Boolean algebras is Givant and Halmos [23] which is blessed with an extensive bibliography. For a description of computer circuits and Boolean algebras very much at the level of this book see Petzold [61] and for a nice account of both Boole and Shannon, the electrical engineer who applied Boolean algebras to circuit design, see Nahin [58].

The history of logic is catered for in a graphic novel [13], a thorough tome [41] and something roughly half-way [12]. For the philosophy of the Stoics

---

[19] You can read more about Euclidean geometry in my book [44].

[20] The name given to logic applied to mathematics is *model theory*.

in particular, try [49], and for an exponent of Stoic philosophy read [3]. The relationship between logic and mathematics is discussed, at an advanced level, in Hart [28] and Hatcher [29].

If you want introductions to the concept of proof in mathematics then Hammack [26] is a good place to start whereas Exner and Rosskopf [15] is interesting in that it uses formal logic to examine proofs in mathematics. I don't use a lot of mathematics in my book, but the two books [44, 46] might be useful for extra reading.

Petzold's other book [62] walks the reader through Turing's famous paper. This paper is unusual in the history of mathematics in that it is has virtually no prerequisites. That doesn't mean it is an easy read, you have to work; but that's all you have to do, rather than require years of mathematical training. The limitations of computers in the work of Turing and others is well explained in [27].

The role of logic in philosophy is important, particularly in the British tradition of analytic philosophy.[21] See Tennant [74] and also Smith [66].[22] The creators of modern logic were by and large interesting people, particularly Russell, who was sent to jail no less than two times.[23] There are a number of biographies worth checking out, such as: [24, 32, 47, 55, 56].

There is a huge choice of second-course logic books though they tend to divide between those intended for CS students and those intended for Mathematics students. Books on logic aimed squarely at CS students are: Ben-Ari [4], Gallier [18] (quite advanced), Huth and Ryan [37] and Schöning [64] whereas books on logic with a more mathematical bent are: Cori and Lascar [10], Hodel [31], Mendelson [52] and van Dalen [78]. Hodel's book [31] is a natural sequel to mine.

Logic and automata theory are intertwined and intermixed. See Boolos, Burgess and Jeffrey [7] for the details. Good accounts of automata theory can be found in the unconventional Floyd and Beigel [17], the standard Kozen [42] and the semigroup-theoretical Lawson [43]. $\mathcal{NP}$-complete problems, and more hints on how to win that million dollars, can be found in the classic book by Garey and Johnson [19] and from a broader perspective in Papadimitriou [60]. The early development of automata theory was bound up with the notion of *neural networks*. The first outing for this concept followed closely on Turing's work [50].

There are many older books on circuit design, such as Mano [48], but this is an area with a connection to a multi-billion dollar industry and so things

---

[21]To be contrasted with *continental philosophy* which is a branch of literature [70].

[22]There is, of course, Wittgenstein's *Tractatus Logico-Philosophicus* [79]. For a lighter take on this opus, try *M.A. Numminen sings Wittgenstein* at https://www.youtube.com/watch?v=57PWqFowq-4 for the original and at https://www.youtube.com/watch?v=CGksgZKecKE for a live performance in Oslo. I discovered an audio tape of Numminen's setting of the *Tractatus* in a now defunct bookshop in Bangor Uchaf in the early 90s.

[23]I don't mean this as approbation; it's just that it's so unexpected.

change fast. A modern book is [11]. For the quantum-mechanical future of computers, perhaps, see [1] and [59].

Kant stated that Euclidean geometry is the only conceivable geometry, but Bolyai, Gauss, and Lobachevsky begged to differ and ushered in non-Euclidean geometries, of which there are many. Are there non-classical logics? Yes. Are they interesting? Yes. The one you are most likely to encounter is *intuitionistic logic*. A good place to get your bearings in this strange new world is van Dalen [78]. A more recent non-classical logic but with a classical flavour is *linear logic* invented by Jean-Yves Girard who found it hidden within Gentzen's sequent calculus. There is also a book [22]. One possible direction for logic in mathematics is explained in [77], which I have included out of sheer devilry.

Finally, it is singularly appropriate to conclude with the work of Kurt Gödel who, along with Alan Turing, is one of the tutelary deities of this book. We have described a version of his completeness theorem for FOL in this chapter. This result formed the basis of his PhD thesis. But he is much more famous for his two *incompleteness* theorems which describe the limitations of the very axiomatic method which is the basis of all mathematics; see Section 3.3, for example. An intuitive description of this work, and its cultural and biographical background, can be found in [24]. Proofs of the incompleteness theorems can be found in [7] and [31].

Solutions to all exercises

1. (a) *C was a knight.* The key point is the following. If you ask a knight what
    he is he will say he is a knight whereas if you ask a knave what he is he
    is obliged to lie and so also say that he is a knight. Thus no one on this
    island can say they are a knave. This means that B is lying and so is a
    knave. Hence C was correct in saying that B lies and so C was a knight.

    (b) *A is a knave, B a knight and C a knave.* If all three were knaves then
    C would be telling the truth which would contradict the fact that he is
    a self-proclaimed knave. Therefore *C* is a knave and there is either one
    knight or there are two knights. Suppose that there were exactly two
    knights. Then these would have to be A and B. But they contradict each
    other. It follows that exactly one of them is a knight. Hence A is a knave
    and B is a knight.

2. *Sam drinks water and Mary owns the aardvark.* The following table shows all
    the information you should have deduced.

	1	2	3	4	5
House	Yellow	Blue	Red	White	Green
Pet	Fox	Horse	Snails	Dog	Aardvark
Name	Sam	Tina	Sarah	Charles	Mary
Drink	Water	Tea	Milk	Orange juice	Coffee
Car	Bentley	Chevy	Oldsmobile	Lotus	Porsche

You should check that the above solution is consistent with all the information
you have been given.

The solution of this problem requires thought processes typical of logic. A
reasonable starting point is the following table where I have entered all possibilities. Observe that I have had to use abbreviations to fit all the information
in.

	1	2	3	4	5
House	R, G, W, Y, B	R, G, W, Y, B	R, G, W, Y, B	R, G, W, Y, B	R, G, W, Y, B
Pet	D, S, F, H, Aa	D, S, F, H, Aa	D, S, F, H, Aa	D, S, F, H, Aa	D, S, F, H, Aa
Name	S, C, T, M, Sam	S, C, T, M, Sam	S, C, T, M, Sam	S, C, T, M, Sam	S, C, T, M, Sam
Drink	T, C, M, OJ, W	T, C, M, OJ, W	T, C, M, OJ, W	T, C, M, OJ, W	T, C, M, OJ, W
Car	O, B, C, L, P	O, B, C, L, P	O, B, C, L, P	O, B, C, L, P	O, B, C, L, P

Using clues (h), (i) and (n), we can make positive entries in three cells and, in
addition, remove those entries from every other cell in the corresponding row.

	1	2	3	4	5
House	R, G, W, Y	**Blue**	R, G, W, Y	R, G, W, Y	R, G, W, Y
Pet	D, S, F, H, Aa	D, S, F, H, Aa	D, S, F, H, Aa	D, S, F, H, Aa	D, S, F, H, Aa
Name	**Sam**	S, C, T, M	S, C, T, M	S, C, T, M	S, C, T, M
Drink	T, C, OJ, W	T, C, OJ, W	**Milk**	T, C, OJ, W	T, C, OJ, W
Car	O, B, C, L, P	O, B, C, L, P	O, B, C, L, P	O, B, C, L, P	O, B, C, L, P

The clues (h), (i) and (n) have now done their job and will play no further role. We now go through the list of remaining clues to determine whether new conclusions can be drawn. Clue (a) tells us that Sarah lives in the red house so she cannot live in the second house because that is blue. It follows that Sam cannot live in the red house. Clue (b) tells us that Charles owns the dog so Sam cannot own the dog. Clue (c) tells us that coffee is drunk in the green house and so coffee cannot be drunk in the blue house. Clue (d) tells us that Tina drinks tea so Sam cannot drink tea and house 3 drinks milk so the owner cannot be Tina. Clue (e) tells us that (white–green) are next to each other. It follows that house 1 cannot be white or green. Thus house 1 is yellow. It follows by clue (c) that coffee is not drunk in house 1. But now clue (g) tells us that Sam owns the Bentley; we have now used up clue (g). Clue (f) tells us that Sam doesn't own the snails. Clue (k) tells us that house 2 owns the horse. Clue (l) tells us that house 3 cannot own the Lotus and house 1 cannot drink OJ. Clue (f) tells us that house 2 cannot own the Oldsmobile.

	1	2	3	4	5
House	**Yellow**	**Blue**	R, W	R, G, W	R, G, W
Pet	F, Aa	**Horse**	D, S, F, Aa	D, S, F, Aa	D, S, F, Aa
Name	**Sam**	T, M	S, C, M	S, C, T, M	S, C, T, M
Drink	**Water**	T, OJ	**Milk**	T, C, OJ	T, C, OJ
Car	**Bentley**	C, L, P	O, C, P	O, C, L, P	O, C, L, P

In house 2 either tea is drunk or OJ. Suppose house 2 drank OJ. Then house 2 would own the Lotus by clue (l) and, since by clue (m) Mary drives the Porsche, it would follow that house 2 was owned by Tina. But by clue (d), we know that Tina drinks tea. This is a contradiction. It follows that in house 2 tea is drunk. Thus by clue (d), Tina lives in house 2. By clue (m), Tina cannot drive the Porsche. By clue (l), Tina cannot drive the Lotus. It follows that Tina drives the Chevy. By clue (j), it follows that neither house 4 nor 5 can own the fox.

	1	2	3	4	5
House	**Yellow**	**Blue**	R, W	R, G, W	R, G, W
Pet	F, Aa	**Horse**	D, S, F, Aa	D, S, Aa	D, S, Aa
Name	**Sam**	**Tina**	S, C, M	S, C, M	S, C, M
Drink	**Water**	**Tea**	**Milk**	C, OJ	C, OJ
Car	**Bentley**	**Chevy**	O, P	O, L, P	O, L, P

Suppose that house 3 were white. Then house 4 would be green by clue (e) and house 5 would be red. By clue (a), Sarah would live in house 5. By clue (b), she cannot own a dog. By clue (c), she cannot drink coffee so she must drink OJ. By clue (l), she must drive the lotus. Thus house 4 drinks coffee.

By clue (f), Sarah cannot own the snails. Thus Sarah must own the aardvark. This leaves either the dog or the snails in house 4. If it is the dog, then by clue (b) Charles lives in house 4. But by clue (m), he cannot own the Porsche and so must own the Oldsmobile but then by clue (f), the Oldsmobile owner has the snails. This is a contradiction. Thus the snails must be in house 4. Since by clue (b), Charles owns the dog, it follows that Mary owns the snails. By clue (f), house 4 owns the Oldsmobile. But by clue (m), Mary owns the Porsche. This is a contradiction. We deduce that house 3 cannot be white, and so it must be red. It follows that house 4 is white and house 5 is green.

	1	2	3	4	5
House	**Yellow**	**Blue**	**Red**	**White**	**Green**
Pet	F, Aa	**Horse**	D, S, F, Aa	D, S, Aa	D, S, Aa
Name	**Sam**	**Tina**	S, C, M	S, C, M	S, C, M
Drink	**Water**	**Tea**	**Milk**	C, OJ	C, OJ
Car	**Bentley**	**Chevy**	O, P	O, L, P	O, L, P

By clue (a), Sarah is in house 3. By clue (b), she cannot own the dog. By clue (c), coffee is drunk in the green house and so OJ is drunk in the white house. By clue (l), the Lotus is driven in house 4. By clue (f), house 4 cannot own snails. By clue (m), Mary drives the Porsche so house 4 must be Charles and by clue (b) the pet is a dog at which point the solution drops out.

3. There are two possible secret numbers consistent with the information given: 2745 or 4725.

4. This difficult question is taken from the famous book [33]. The answer to the question is no. The key is to focus on the number of I's in a string which we call the I-count. Rule-I does not change the I-count. Rule-II doubles the I-count. Rule-III reduces the I-count by 3. Rule-IV does not change the I-count. We begin with a string whose I-count is 1 and our goal is to obtain a string whose I-count is 0. The problem reduces to showing that applying the above rules to a string whose I-count is 1 never results in a string whose I-count is 0. This amounts to showing that if 3 does not divide n then 3 does not divide $2n$, and if 3 does not divide n then 3 does not divide $n - 3$.

5. Here is the completed puzzle taken from *Solving Sudoku* by Michael Mepham available at www.sudoku.org.uk/PDF/Solving_Sudoku.pdf.

8	1	3	4	2	9	7	6	5
4	6	2	5	7	1	8	3	9
7	9	5	3	6	8	1	4	2
2	4	7	1	5	3	9	8	6
5	3	9	8	4	6	2	1	7
6	8	1	2	9	7	4	5	3
9	7	8	6	1	5	3	2	4
1	2	6	7	3	4	5	9	8
3	5	4	9	8	2	6	7	1

Exercises 1.1

1.

(a) T.

(b) F.

(c) T.

(d) F.

(e) T.

(f) T.

(g) F.

(h) T.

(i) T.

(j) T.

2. The following statement is supposed to be true

> 'a card has a vowel on one side $\rightarrow$ it has an even number on the other side'.

You would falsify this statement if you could find a card that had a vowel on one side but did not have an even number on the other. Thus, clearly, I need to turn over the first card to check that it has an even number on the other side. The next two cards play no role. Now look at the last card. If it had a vowel on the other side then it would falsify the statement. So, I need to turn that card over as well. Thus I only need to turn over two cards.

It is possible to answer the question using just PL though it really needs the notion of logical equivalence described in Section 1.4. Observe that the truth tables for $p \rightarrow q$ and $\neg q \rightarrow \neg p$ are the same. Thus these wff have the same meaning. The first wff immediately tells us to turn over the first card. The second wff says that 'a card had an odd number on one side $\rightarrow$ it has a consonant on the other'. Thus we need to turn over the fourth card to check that it does show a consonant.

This question has its origins in the psychology of reasoning. Read the wikipedia article if you want to know more. What is interesting is that if the same problem is couched in the language of social rules it becomes a lot easier to solve. Here is an example. You have to be 18 or over to buy alcohol in the UK. Below is some data about four people: on one side is their age and on the other what they want to buy. Who do you have to check?

My guess is that you *immediately* saw that you had to check the age of the guy buying beer and the 16-year-old to check that they were not trying to buy booze. But the logical structure of this problem is identical to the one originally posed.

Exercises 1.2

1.

(a)

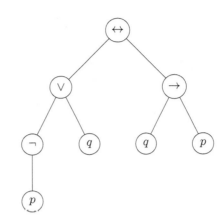

(b)

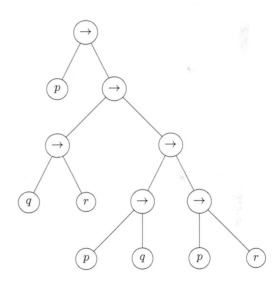

(c)

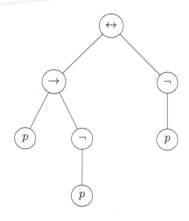

(d)

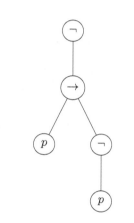

(e)

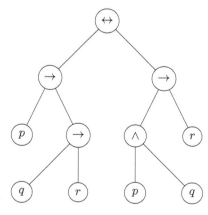

2. $A = B$ because the order that elements are listed in a set plays no role. $B = C$ because neither order nor repetitions play a role in what a set is.

3. (a) 0.

   (b) 1.

   (c) 2.

   (d) 3.

Exercises 1.3

1.

   (a) This is always satisfiable except for one truth assignment $p = T$ and $q = F$ leading to an F.

   (b) This is always satisfiable except for one truth assignment $p = T$ and $q = F$ leading to an F.

   (c) This is satisfiable with the satisfying truth assignments being $p = T, q = T, r = T$ and $p = T, q = T, r = F$ and $p = F, q = F, r = F$.

2.

   (a) This is not a tautology.

   (b) This is a tautology.

   (c) This is a tautology.

   (d) This is not a tautology.

   (e) This is a tautology.

3. There are potentially many different solutions to this question.

Column	wff
1	$(p \lor \neg p) \land (q \lor \neg q)$
2	$p \lor q$
3	$p \lor \neg q$
4	p
5	$p \to q$
6	q
7	$p \leftrightarrow q$
8	$p \land q$
9	$\neg(p \land q)$
10	$p \oplus q$
11	$\neg q$
12	$\neg(p \to q)$
13	$\neg p$
14	$\neg p \land q$
15	$\neg(p \lor q)$
16	$(p \land \neg p) \land q$

Exercises 1.4

1. To show that $A \equiv B$ construct truth tables for A and B separately and show that they are the same.

2.   (a) $p \vee \neg p$ is a tautology.

      (b) $p \wedge \neg p$ is a contradiction.

      (c)   Clear.

      (d)   Clear.

      (e)   Clear.

      (f)   Clear.

      (g)   If p is true then $p \to f$ is false. If p is false then $p \to f$ is true. It follows that $p \to f \equiv \neg p$.

      (h)   If p is true then $t \to p$ is true. If p is false then $t \to p$ is false. It follows that $t \to p \equiv p$.

3. Check that both $(p \oplus q) \oplus r$ and $p \oplus (q \oplus r)$ have the following truth table.

p	q	r	
T	T	T	T
T	T	F	F
T	F	T	F
T	F	F	T
F	T	T	F
F	T	F	T
F	F	T	T
F	F	F	F

4. It is important to lay out the solutions clearly with proper annotations.

      (a)

$$
\begin{aligned}
(p \to q) \wedge (p \vee q) &\equiv (\neg p \vee q) \wedge (p \vee q) \text{ since } x \to y \equiv \neg x \vee y \\
&\equiv (\neg p \wedge p) \vee q \text{ by distributivity} \\
&\equiv f \vee q \text{ since } \neg p \wedge p \text{ is a contradiction} \\
&\equiv q.
\end{aligned}
$$

      (b)

$$
\begin{aligned}
(p \to r) \vee (q \to r) &\equiv (\neg p \vee r) \vee (\neg q \vee r) \text{ since } x \to y \equiv \neg x \vee y \\
&\equiv \neg p \vee r \vee \neg q \vee r \text{ by associativity} \\
&\equiv \neg p \vee \neg q \vee r \vee r \text{ by commutativity} \\
&\equiv \neg p \vee \neg q \vee r \text{ by idempotence} \\
&\equiv \neg(p \wedge q) \vee r \text{ by De Morgan} \\
&\equiv (p \wedge q) \to r \text{ since } x \to y \equiv \neg x \vee y.
\end{aligned}
$$

      (c)

$$
\begin{aligned}
(p \to q) \vee (p \to r) &\equiv (\neg p \vee q) \vee (\neg p \vee r) \text{ since } x \to y \equiv \neg x \vee y \\
&\equiv \neg p \vee \neg p \vee q \vee r \text{ by associativity and commutativity} \\
&\equiv \neg p \vee q \vee r \text{ by idempotence} \\
&\equiv p \to (q \vee r) \text{ since } x \to y \equiv \neg x \vee y.
\end{aligned}
$$

5. (a) $A(p, q, r, s)$ is true when exactly one of p, q, r, s is true.

    (b) Define

    $$A(p_1, \ldots, p_n) = \left(\bigvee_{i=1}^{n} p_i \right) \wedge \left(\bigwedge_{1 \le k < l \le n} \neg(p_k \wedge p_l) \right).$$

    We claim that $A(p_1, \ldots, p_n)$ is true when exactly one of the p_i is true. Suppose first, that p_s is true and all other atoms are false. The first bracket is clearly true. Now look at the second bracket. Then $p_k \wedge p_l$ where $k \ne l$ is either $p_s \wedge p_l$ where $l \ne s$ or $p_k \wedge p_l$ where $k \ne l \ne s$. Therefore, under the specific truth assignment above, $p_k \wedge p_l$ is always false and so the second bracket is true. It follows that $A(p_1, \ldots, p_n)$ is true. Now suppose that $A(p_1, \ldots, p_n)$ is true. Then both brackets have to be true. It follows that at least one p_i is true. Suppose that at least two were true: p_s and p_t where $s < t$. Then the second bracket would contain the wff $\neg(p_s \wedge p_t)$ which is false giving us a contradiction. It follows that exactly one p_i is true.

    I will use the notation $\text{xor}(p_1, \ldots, p_n)$ instead of $A(p_1, \ldots, p_n)$.

6. (a) The wff $\neg A$ is logically equivalent to one in which $\vee$ is swapped with $\wedge$, and $\wedge$ is swapped with $\vee$ and where each atom p is replaced by $\neg p$. But we only want to interchange $\vee$ and $\wedge$. Thus we replace each atom in p in A by $\neg p$ and then use double negation. For example, let $A(p, q, r) = p \vee (q \wedge \neg r)$. Then $\neg A(\neg p, \neg q, \neg r) \equiv p \wedge (q \vee \neg r) = A^*$.

    (b) By symmetry, it is enough to show that $\vDash A \leftrightarrow B$ implies that $\vDash A^* \leftrightarrow B^*$. Let the atoms in A and B be $p_1, \ldots, p_n$. Suppose that $A^* \leftrightarrow B^*$ is not a tautology. Then there is some assignment of truth values to the atoms such that A^* is true and B^* is false, or vice versa. But by part (a), we have that $\neg A^* = A(\neg p_1, \ldots, \neg p_n)$ and $\neg B^* = B(\neg p_1, \ldots, \neg p_n)$. Thus $A(\neg p_1, \ldots, \neg p_n)$ is false and $B(\neg p_1, \ldots, \neg p_n)$ is true. By reversing the truth values assigned to $p_1, \ldots, p_n$ it follows that A is false and B is true but this contradicts the fact that A and B assume the same truth values for the same truth assignments to the atoms.

Exercises 1.5

1. (a) The wff A is the following

    $$(p \oplus q) \quad \wedge \quad (r \oplus s) \wedge (t \oplus u) \wedge (v \oplus w) \wedge (p \oplus r) \wedge (q \oplus s)$$
    $$\wedge \quad (t \oplus v) \wedge (u \oplus w) \wedge (p \oplus t) \wedge (r \oplus v) \oplus (q \oplus u) \wedge (s \oplus w).$$

    (b) There are $2^8 = 256$ rows of the truth table.

    (c) 2.

    (d) The atoms assigned the value T will tell us which numbers occur in which cells.

    (e) The rows of the truth table for A that have the output T are as follows

p	q	r	s	t	u	v	w
T	F	F	T	F	T	T	F
F	T	T	F	T	F	F	T

These two rows of the truth table correspond to the following two solutions of the Sudoku.

1	2	2	1
2	1	1	2

2. (a) $I = p_{1,2,1} \wedge p_{1,4,4} \wedge p_{1,5,2} \wedge \ldots \wedge p_{9,8,7}$. There are 29 atoms in I in total.

(b) Fix i and j. Then $E_{ij} = \text{xor}(p_{i,j,1}, \ldots, p_{i,j,9})$.

(c) $E = \bigwedge_{1 \le i,j \le 9} E_{i,j}$. Thus E is true when every cell contains exactly one digit.

(d) Fix i and k. Then $\text{xor}(p_{i,1,k}, p_{i,2,k}, \ldots, p_{i,9,k})$ is true when the digit k occurs exactly once in row i. Thus

$$R_i = \bigwedge_{k=1}^{9} \text{xor}(p_{i,1,k}, p_{i,2,k}, \ldots, p_{i,9,k})$$

is true when each digit occurs exactly once in row i.

(e) $R = \bigwedge_{i=1}^{9} R_i$ is true when each row contains each digit exactly once.

(f) This is similar to what we did in the above two parts. Put $C = \bigwedge_{j=1}^{9} C_j$ where

$$C_j = \bigwedge_{k=1}^{9} \text{xor}(p_{1,j,k}, \ldots, p_{9,j,k}).$$

(g) We look at block 5 where $4 \le i \le 6$ and $4 \le j \le 6$. Then

$$W_5 = \bigwedge_{k=1}^{9} \text{xor}(p_{4,4,k}, p_{4,5,k}, p_{4,6,k}, \ldots, p_{6,6,k})$$

and W_5 is true when each digit occurs exactly once in block 5. In a similar way, we may define W_l to be true when each digit occurs exactly once in block l.

(h) $W = \bigwedge_{l=1}^{9} W_l$.

(i) We now consider when P is true. Suppose that you come up with a potential solution to this Sudoku. This means that for each cell c_{ij} you enter a digit k. This assigns the value true to the atom $p_{i,j,k}$. Once you have done this for every cell, all the remaining atoms are assigned the value false. Thus a potential solution to the puzzle leads to a truth assignment to all 729 atoms. With respect to this truth assignment, P is either true or false. It will be false if any one of the constraints is violated. If none of the constraints is violated, then your potential solution is correct. Thus a correct solution to the puzzle leads to a truth assignment to the atoms that satisfies P.

Now suppose that you have found a truth assignment that satisfies P. If $P_{i,j,k}$ is true then put digit k in cell c_{ij}. I claim that because of the way that P is defined, this will lead to every empty cell being assigned a digit. This is because if P is true then E is true and so each cell contains exactly one digit. Thus the truth assignment leads to a potential solution to the puzzle. But since P is true all constraints are satisfied and so the Sudoku has, in fact, been solved.

3. Define A to be the conjunction of the following wff.

- Initial state: $q(0) = 1$.
- Input constraints:

$$((i(0) = a) \oplus (i(0) = b)) \wedge ((i(1) = a) \oplus (i(1) = b)).$$

- Unique state at each instant of time $t = 0$, $t = 1$ and $t = 2$:

$$\text{xor}\,((q(0) = 1), (q(0) = 2))$$
$$\wedge \quad \text{xor}\,((q(1) = 1), (q(1) = 2))$$
$$\wedge \quad \text{xor}\,((q(2) = 1), (q(2) = 2))\,.$$

- First set of possible state transitions:

$$(q(0) = 1) \wedge (i(0) = a) \rightarrow (q(1) = 1)$$
$$\wedge \quad (q(0) = 1) \wedge (i(0) = b) \rightarrow (q(1) = 2)$$
$$\wedge \quad (q(0) = 2) \wedge (i(0) = a) \rightarrow (q(1) = 1)$$
$$\wedge \quad (q(0) = 2) \wedge (i(0) = b) \rightarrow (q(1) = 2).$$

- Second set of possible state transitions:

$$(q(1) = 1) \wedge (i(1) = a) \rightarrow (q(2) = 1)$$
$$\wedge \quad (q(1) = 1) \wedge (i(1) = b) \rightarrow (q(2) = 2)$$
$$\wedge \quad (q(1) = 2) \wedge (i(1) = a) \rightarrow (q(2) = 1)$$
$$\wedge \quad (q(1) = 2) \wedge (i(1) = b) \rightarrow (q(2) = 2).$$

- Finally, there is the actual input string of length 2.

$$(i(0) = a) \wedge (i(1) = b).$$

We look at the circumstances under which A is true. '$q(0) = 1$' is true and so '$q(0) = 2$' is false; '$i(0) = a$' is true and so '$i(0) = b$' is false. We therefore deduce from the first set of state transitions that '$q(1) = 1$' is true and so '$q(1) = 2$' is false. From the second set of state transitions we deduce that '$q(2) = 2$' is true and so '$q(2) = 1$' is false.

Exercises 1.6

1.

$\neg p$	$p \text{ nand } p$	$p \text{ nor } p$
$p \wedge q$	$(p \text{ nand } q) \text{ nand } (p \text{ nand } q)$	$(p \text{ nor } p) \text{ nor } (q \text{ nor } q)$
$p \vee q$	$(p \text{ nand } p) \text{ nand } (q \text{ nand } q)$	$(p \text{ nor } q) \text{ nor } (p \text{ nor } q)$
$p \rightarrow q$	$p \text{ nand } (q \text{ nand } q)$	$((p \text{ nor } p) \text{ nor } q) \text{ nor } ((p \text{ nor } p) \text{ nor } q)$

2. (a) We have that $p \rightarrow q \equiv \neg p \vee q$. It follows that $p \vee q \equiv \neg p \rightarrow q$. Thus from $\neg$ and $\rightarrow$ we can construct $\vee$, but $\vee$ and $\neg$ together are adequate, and so $\neg$ and $\rightarrow$ are adequate.

   (b) Observe that $p \rightarrow \boldsymbol{f} \equiv \neg p$. Adequacy now follows from part (a).

3. (a) If $T * T = T$ we would never be able to construct negation using $*$.

   (b) If $F * F = F$ we would never be able to construct negation using $*$.

   (c) The possible truth tables for $*$ are therefore as follows:

x	y	nand	nand	$\neg y$	$\neg x$
T	T	F	F	F	F
T	F	T	F	T	F
F	T	T	F	F	T
F	F	T	T	T	T

   If we look at the third column, the value of $x_1 * \ldots * x_n$ will either be x_n or $\neg x_n$. It will therefore be true for exactly half the truth assignments. But there are plenty of wff which are not true for exactly half the truth assignments. Therefore it cannot be adequate on its own. A similar argument applies to the fourth column.

Exercises 1.7

1. (a) $p \wedge \neg q \wedge r$.

   (b) $(p \wedge q \wedge \neg r) \vee (p \wedge \neg q \wedge \neg r)$.

   (c) $(p \wedge q \wedge r) \vee (p \wedge \neg q \wedge r) \vee (\neg p \wedge q \wedge r) \vee (\neg p \wedge \neg q \wedge r)$.

2. I shall prove the 'hard' direction. Suppose that

$$\{\{a\}, \{a, b\}\} = \{\{c\}, \{c, d\}\}.$$

   Then $\{a\} = \{c\}$ or $\{a\} = \{c, d\}$. If the latter then $a = c = d$ but then $\{a, b\} = \{a\}$ and so $a = b = c = d$. Suppose the former. Then $a = c$. There are now two possibilities either $\{a, b\} = \{c, d\}$, in which case $b = d$, or $\{a, b\} = \{c\}$, in which case $a = b = c$ and so $a = d$. It follows that in all cases: $a = c$ and $b = d$.

3. Define an ordered triple (a, b, c) to be $((a, b), c)$. Define an ordered 4-tuple (a, b, c, d) to be $((a, b, c), d)$, and so on.

Exercises 1.8

1. (a)

$$
\begin{aligned}
(p \to q) \to p &\equiv (\neg p \vee q) \to p \\
&\equiv \neg(\neg p \vee q) \vee p \\
&\equiv (\neg\neg p \wedge \neg q) \vee p \\
&\equiv (p \wedge \neg q) \vee p \text{ NNF and DNF.}
\end{aligned}
$$

(b)

$$
p \to (q \to p) \equiv \neg p \vee \neg q \vee p \text{ NNF and DNF.}
$$

(c)

$$
\begin{aligned}
(q \wedge \neg p) \to p &\equiv \neg(q \wedge \neg p) \vee p \\
&\equiv (\neg q \vee \neg\neg p) \vee p \\
&\equiv \neg q \vee p \vee p \\
&\equiv \neg q \vee p \text{ NNF and DNF.}
\end{aligned}
$$

(d) $(p \vee q) \wedge r$ is already in NNF and $(p \vee q) \wedge r \equiv (p \wedge r) \vee (q \wedge r)$ is in DNF.

(e) $p \to (q \wedge r) \equiv \neg p \vee (q \wedge r)$ is in NNF and DNF.

(f) $(p \vee q) \wedge (r \to s) \equiv (p \vee q) \wedge (\neg r \vee s)$ is in NNF and $(p \vee q) \wedge (\neg r \vee s) \equiv (p \wedge \neg r) \vee (p \wedge s) \vee (q \wedge \neg r) \vee (q \wedge s)$ is in DNF.

2. We use the calculations from Question 1 above.

   (a) $(p \wedge \neg q) \vee p \equiv (p \vee p) \wedge (\neg q \vee p) \equiv p \wedge (\neg q \vee p)$.

   (b) $(\neg p \vee \neg q \vee p)$.

   (c) $(\neg q \vee p)$.

   (d) $(p \vee q) \wedge (r)$.

   (e) $\neg p \vee (q \wedge r) \equiv (\neg p \vee q) \wedge (\neg p \vee r)$.

   (f) $(p \vee q) \wedge (\neg r \vee s)$.

3. (a)

p	q	r	A
T	T	T	T
T	T	F	F
T	F	T	T
T	F	F	F
F	T	T	T
F	T	F	F
F	F	T	T
F	F	F	F

   (b) $(p \wedge q \wedge r) \vee (p \wedge \neg q \wedge r) \vee (\neg p \wedge q \wedge r) \vee (\neg p \wedge \neg q \wedge r)$.

(c)

p	q	r	$\neg A$
T	T	T	F
T	T	F	T
T	F	T	F
T	F	F	T
F	T	T	F
F	T	F	T
F	F	T	F
F	F	F	T

(d) $(p \wedge q \wedge \neg r) \vee (p \wedge \neg q \wedge \neg r) \vee (\neg p \wedge q \wedge \neg r) \vee (\neg p \wedge \neg q \wedge \neg r)$.

(e) $(\neg p \vee \neg q \vee r) \wedge (\neg p \vee q \vee r) \wedge (p \vee \neg q \vee r) \wedge (p \vee q \vee r)$.

4. (a)

$$
\begin{aligned}
((p \wedge q) \to r) \wedge (\neg(p \wedge q) \to r) &\equiv (\neg(p \wedge q) \vee r) \wedge (\neg\neg(p \wedge q) \vee r) \\
&\equiv (\neg p \vee \neg q \vee r) \wedge ((p \wedge q) \vee r).
\end{aligned}
$$

(b) $(\neg p \wedge p \wedge q) \vee (\neg p \wedge r) \vee (\neg q \wedge q \wedge p) \vee (\neg q \wedge r) \vee (r \wedge p \wedge q) \vee r$.

(c) $(\neg p \vee \neg q \vee r) \wedge (p \vee r) \wedge (q \vee r)$.

(d)

$$
\begin{aligned}
A &\equiv ((p \wedge q) \to r) \wedge (\neg(p \wedge q) \to r) \\
&\equiv (\neg(p \wedge q) \vee r) \wedge ((p \wedge q) \vee r) \\
&\equiv (\neg(p \wedge q) \wedge (p \wedge q)) \vee r \\
&\equiv r.
\end{aligned}
$$

5. $(\neg p \vee ((\neg q \vee r) \wedge (\neg r \vee q))) \wedge (p \vee (q \wedge \neg r) \vee (r \wedge \neg q))$. This question illustrates the fact that although we can write any wff in NNF, there are often good reasons for not always doing so.

6. (a) $(q \to p) \wedge (r \to q)$. Satisfiable with all atoms taking the value F.

(b) $((q \wedge r) \to p) \wedge ((s \wedge u) \to f) \wedge ((p \wedge q) \to r) \wedge (t \to p) \wedge (t \to q)$. Satisfiable with the following truth assignment

p	q	r	s	u
T	T	T	F	F

(c) $(t \to p) \wedge (t \to q) \wedge ((p \wedge q) \to f) \wedge (p \to r)$. Unsatisfiable

7. The truth table for $p \oplus q$ has output F when both p and q are assigned the truth value F. It follows that if $p \oplus q$ is logically equivalent to a Horn formula, that Horn formula must contain a wff of the form $t \to r$ for some atom r. But then any truth assignment making $p \oplus q$ true must have the atom p, at least, in common taking the value T. However, $p \oplus q$ is true precisely when p and q take different truth values.

Exercises 1.9

1. (a) A says that you should 2-colour the graph so that adjacent vertices have different colours.

   (b) It is a contradiction.

   (c) It is impossible to 2-colour this graph in this way.

Exercises 1.10

1. The following can all be checked using the truth table generator [76].

   (a) $(p \rightarrow q) \rightarrow (\neg q \vee \neg p)$ is not a tautology.

   (b) Show that $\vDash ((p \rightarrow q) \wedge (\neg q \rightarrow p)) \rightarrow q$.

   (c) Show that $\vDash ((p \rightarrow q) \wedge (r \rightarrow s) \wedge (p \vee r)) \rightarrow (q \vee r)$.

   (d) Show that $\vDash ((p \rightarrow q) \wedge (r \rightarrow s) \wedge (\neg q \vee \neg s)) \rightarrow (\neg p \vee \neg r)$.

2. Suppose that $x_1 \vee x_2 \vee x_3$ and $\neg x_1 \vee y_2 \vee y_3$ are both true. There are two cases. Suppose that x_1 is true, then $\neg x_1$ is false. Thus at least one of y_2 or y_3 is true and so $x_2 \vee x_3 \vee y_2 \vee y_3$ is true. Suppose that x_1 is false, then at least one of x_2 or x_3 is true and so $x_2 \vee x_3 \vee y_2 \vee y_3$ is true.

3. The proofs of (a), (b) and (c) are all straightforward.

   (d) Suppose that $A_1, \ldots, A_m, X \vDash B_1, \ldots, B_n$ is a valid argument. We prove that $A_1, \ldots, A_m \vDash \neg X, B_1, \ldots, B_n$ is a valid argument. Suppose that $A_1, \ldots, A_m$ are all true. If X is false then $\neg X$ is true and we are done. If X is true then $B_1 \vee \ldots \vee B_n$ is true and we are done.

   The proof of (e) is straightforward.

   (f) Suppose that $A_1, \ldots, A_m$ are all true, then $B_1 \vee \ldots \vee B_n \vee X$ is true and $B_1 \vee \ldots \vee B_n \vee Y$. Thus, by distributivity, $B_1 \vee \ldots \vee B_n \vee (X \wedge Y)$ is true.

   The proof of (g) is straightforward.

   The proof of (h) follows by (e) and (g).

Exercises 1.11

1. (a)

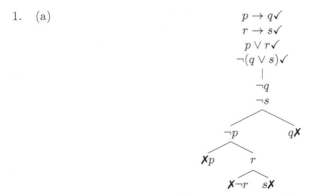

All branches close, so the argument is valid.

(b)

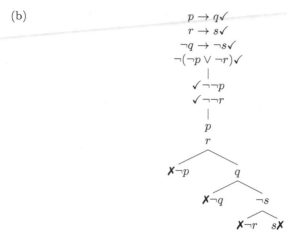

All branches close, so the argument is valid.

(c)

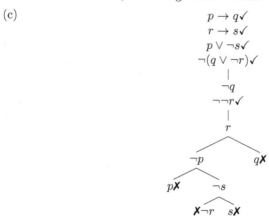

All branches close, so the argument is valid.

2. (a)

$$\neg(q \to (p \to q))\checkmark$$
$$|$$
$$q$$
$$\neg(p \to q)\checkmark$$
$$|$$
$$p$$
$$\neg q\mathsf{X}$$

The truth tree closes, and so this is a tautology.

(b)

$$\neg((p \to q) \wedge (q \to r)) \to (p \to r))\checkmark$$
$$|$$
$$(p \to q) \wedge (q \to r)\checkmark$$
$$\neg(p \to r)\checkmark$$
$$|$$
$$p \to q\checkmark$$
$$q \to r\checkmark$$
$$|$$
$$p$$
$$\neg r$$

The truth tree closes, and so this is a tautology.

(c)

$$\neg(((p \to q) \wedge (p \to r)) \to (p \to (q \wedge r)))\checkmark$$
$$|$$
$$(p \to q) \wedge (p \to r)\checkmark$$
$$\neg(p \to (q \wedge r))\checkmark$$
$$|$$
$$p \to q\checkmark$$
$$p \to r\checkmark$$
$$|$$
$$p$$
$$\neg(q \wedge r)\checkmark$$

The truth tree closes, and so this is a tautology.

(d)

$$\neg(((p \to r) \wedge (q \to r)) \wedge (p \vee q)) \to r)\checkmark$$
$$|$$
$$((p \to) \wedge (q \to r)) \wedge (p \vee q)\checkmark$$
$$\neg r$$
$$|$$
$$p \to r\checkmark$$
$$q \to r\checkmark$$
$$p \vee q\checkmark$$

The truth tree closes, and so this is a tautology.

3. Pooh's argument in symbolic form is

$$p, p \to q, q \to r, r \to s, s \to t \vDash t.$$

The truth tree for this argument is as follows.

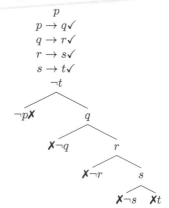

Since the tree closes, the argument is valid.

Exercises 1.12

1. (a)

$$\frac{\dfrac{\neg q, p \Rightarrow p}{\neg q \Rightarrow \neg p, p} \quad \dfrac{q \Rightarrow q, \neg p}{\neg q, q \Rightarrow \neg p}}{p \rightarrow q, \neg q \Rightarrow \neg p}$$

Since the leaves are all axioms, this means that the argument is valid.

(b)

$$\frac{\dfrac{\dfrac{r, p, q \Rightarrow r}{q \rightarrow r, p, q \Rightarrow r} \quad p, q \Rightarrow r, q}{p \rightarrow q, q \rightarrow r, p \Rightarrow r} \quad q \rightarrow r, p \Rightarrow r, p}{p \rightarrow q, q \rightarrow r \Rightarrow p \rightarrow r}$$

Since the leaves are all axioms, this means that the argument is valid.

(c)

$$\frac{\dfrac{r \rightarrow s, p \vee r, q \Rightarrow q, s \quad \dfrac{r \rightarrow s, p \Rightarrow q, s, p \quad \dfrac{s, r \Rightarrow q, s, p \quad r \Rightarrow q, s, p, r}{r \rightarrow s, r \Rightarrow q, s, p}}{r \rightarrow s, p \vee r \Rightarrow q, s, p}}{\dfrac{p \rightarrow q, r \rightarrow s, p \vee r \Rightarrow q, s}{p \rightarrow q, r \rightarrow s, p \vee r \Rightarrow q \vee s}}$$

Since the leaves are all axioms, this means that the argument is valid.

2. (a)

$$\frac{\dfrac{p \Rightarrow p}{\Rightarrow p, \neg p}}{\Rightarrow p \vee \neg p}$$

Since the leaves are all axioms, this means the wff is a tautology.

(b)

$$\frac{\dfrac{p, q \Rightarrow p}{p \Rightarrow q \to p}}{\Rightarrow p \to (q \to p)}$$

Since the leaves are all axioms, this means the wff is a tautology.

(c) Observe that

$$\frac{\dfrac{p \to q, q \to r \Rightarrow p \to r}{(p \to q) \wedge (q \to r) \Rightarrow p \to r}}{\Rightarrow ((p \to q) \wedge (q \to r)) \to (p \to r)}$$

But we already gave a proof of the top sequent in part (b) of Question 1. This means that the wff is a tautology.

3. I shall give two sample proofs for each of the two kinds of rule.

We begin with Conjunction (1). Let τ be a valuation so that all the wff $\mathbf{U}, X, Y$ are true and all the wff in $\mathbf{V}$ are false. Then, clearly, τ makes all the wff in $\mathbf{U}, X \wedge Y$ true and all the wff in $\mathbf{V}$ false. It's clear that this argument can be reversed.

Now consider Implication (2). Suppose first that τ is a valuation that makes all the wff in $\mathbf{U}$ true and all the wff in $\mathbf{V}, X$ false. In particular, $\tau(X) = F$ and so from the definition of implication $\tau(X \to Y) = T$. Thus τ makes all the wff in $\mathbf{U}, X \to Y$ true and all the wff in $\mathbf{V}$ false. Suppose now that τ is a valuation that makes all the wff in $\mathbf{U}, Y$ true and all the wff in $\mathbf{V}$ false. In particular, $\tau(Y) = T$ and so $\tau(X \to Y) = T$. Thus τ makes all the wff in $\mathbf{U}, X \to Y$ true and all the wff in $\mathbf{V}$ false.

To complete the argument, we now have to go in the other direction. Let τ be a valuation that makes all the wff in $\mathbf{U}, X \to Y$ true and all the wff in $\mathbf{V}$ false. If $\tau(X \to Y) = T$ then either $\tau(X) = F$ or $\tau(Y) = T$. This now leads to the two possible paths for τ.

4. Let τ be a truth assignment that satisfies $\mathbf{U} \Rightarrow \mathbf{V}, X$ and $X, \mathbf{W} \Rightarrow \mathbf{Z}$. We prove that τ satisfies $\mathbf{U}, \mathbf{W} \Rightarrow \mathbf{V}, \mathbf{Z}$. By assumption, τ makes all the wff in $\mathbf{U}$ true and all the wff in $\mathbf{W}, X$ true. Thus τ makes all the wff in $\mathbf{U}, \mathbf{W}$ true. It follows that τ makes at least one of the wff in $\mathbf{Z}$ true and so it makes at least one of the wff in $\mathbf{V}, \mathbf{Z}$ true.

Exercises 2.1

1. (a) $A = \{2, 4, 6, 8, 10\}$.
   (b) $B = \{1, 3, 5, 7, 9\}$.
   (c) $C = \{6, 7, 8, 9, 10\}$.
   (d) $D = \emptyset$.
   (e) $E = \{2, 3, 5, 7\}$.
   (f) $F = \{1, 2, 3, 4, 7, 8, 9, 10\}$.

2. This follows by Question 3 of Exercises 1.4.

3. This follows from the fact that $p \wedge \neg q \wedge \neg r \equiv p \wedge \neg(q \vee r)$.

4. This follows from the fact that $p \wedge \neg(q \wedge \neg r) \equiv (p \wedge \neg q) \vee (p \wedge r)$.

Exercises 2.2

1. (1) $A \cap B \cap C$.
   (2) $A \cap B \cap \overline{C}$.
   (3) $A \cap \overline{B} \cap C$.
   (4) $\overline{A} \cap B \cap C$.
   (5) $\overline{A} \cap \overline{B} \cap C$.
   (6) $A \cap \overline{B} \cap \overline{C}$.
   (7) $\overline{A} \cap B \cap \overline{C}$.
   (8) $\overline{(A \cup B \cup C)}$.

2. There are sixteen such subsets. These consist of $\emptyset$ and $\{a, b, c, d\}$; four subsets containing exactly one element each; six subsets that contain exactly two elements each; four subsets that contain exactly three elements each.

3. Straightforward.

4. Use truth tables.

5. Each of the sets in Question 3 is defined using the corresponding wff in Question 4. Logical equivalence of the wff translates into equality of the corresponding sets.

6.

$$
\begin{aligned}
a &= a + 0 \text{ by (B3)} \\
&= a + (a\bar{a}) \text{ by (B10)} \\
&= (a + a)(a + \bar{a}) \text{ by (B8)} \\
&= (a + a)1 \text{ by (B9)} \\
&= a + a \text{ by (B6)}.
\end{aligned}
$$

7.

$$
\begin{aligned}
0a &= (a\bar{a})a \text{ by (B10)} \\
&= (\bar{a}a)a \text{ by (B5)} \\
&= \bar{a}a^2 \text{ by (B5) and (B4)} \\
&= \bar{a}a \text{ since } a^2 = a \\
&= 0 \text{ by (B10)}.
\end{aligned}
$$

8.

$$
\begin{aligned}
1 + a \;&=\; (a + \bar{a}) + a \text{ by (B9)} \\
&=\; (\bar{a} + a) + a \text{ by (B2)} \\
&=\; \bar{a} + (a + a) \text{ by (B1)} \\
&=\; \bar{a} + a \text{ since } a + a = a \\
&=\; 1 \text{ by (B9).}
\end{aligned}
$$

9. (a)

$$
\begin{aligned}
a(a + b) \;&=\; a^2 + ab \text{ by distributivity} \\
&=\; a + ab \text{ since } a^2 = a \\
&=\; a \text{ by absorption.}
\end{aligned}
$$

(b)

$$
\begin{aligned}
a(\bar{a} + b) \;&=\; a\bar{a} + ab \text{ by distributivity} \\
&=\; 0 + ab \text{ since } a\bar{a} = 0 \\
&=\; ab.
\end{aligned}
$$

10. (a) x.

   (b) 0.

   (c) $x + y$.

   (d) x.

   (e) $(x + z)\bar{y}$.

   (f) $x + y$.

11. Here is the idea. Details are omitted. Observe that each number in the set B can be written uniquely in the form $2^x 3^y 5^z$ where $x, y, z \in \{0, 1\}$. Thus each number $n = 2^x 3^y 5^z$ in B can be mapped to the subset Y of $\{2, 3, 5\}$ where Y contains 2 if and only if $x = 1$, contains 3 if and only if $y = 1$, and contains 5 if and only if $z = 1$.

12. We deal with the operation $+$ first. By axioms (B2) and (B3), we have that $0+0 = 0$, $1+0 = 1$ and $0+1 = 1$. By Question 6, we have that $1+1 = 1$. Now we deal with the operation $\cdot$. By Question 7 and (B5), we have that $0 \cdot 0 = 0$, $0 \cdot 1 = 0$ and $1 \cdot 0 = 0$. By part (1) of Proposition 2.2.4, we have that $1 \cdot 1 = 1$. Finally, we deal with complementation. We use Proposition 2.2.7. By (B9) and (B10), we have that $0 + \bar{0} = 1$ and $0 \cdot \bar{0} = 0$. But 1 also satisfies both of these equations in place of $\bar{0}$ and so we deduce that $\bar{0} = 1$. Similarly, $1 + \bar{1} = 1$ and $1 \cdot \bar{1} = 0$ implies that $\bar{1} = 0$.

13. We calculate the left-hand side. By De Morgan and double complements, we have that

$$
\overline{(\bar{x} + y)} + y = (x \cdot \bar{y}) + y.
$$

But by distributivity, this is equal to $(x + y) \cdot (\bar{y} + y)$ which is just $x + y$. By symmetry, this is also equal to the right-hand side.

14. See [35, 36].

Exercises 2.3

1. (a) $\bar{x}y\bar{z}$.

   (b) $xyz + \bar{x}\,\bar{y}\,\bar{z}$.

   (c) $xy\bar{z} + x\bar{y}z + \bar{x}yz$.

2. (a) $u = x$.

   (b) $u = x$.

   (c) $u = x + y$.

3. $u = 1$.

4. $u = \bar{x} + \bar{y} + \bar{z}$.

5. (a) 1010.

   (b) 101010.

   (c) 10011001.

   (d) 11111010001.

6. (a) 7.

   (b) 85.

   (c) 455.

7. (a) 110.

   (b) 100011010.

   (c) 111010.

8. A transistor is defined by the following Boolean expression $x \,\square\, y = x \cdot \bar{y}$. Now, $x \text{ nor } y = \overline{(x + y)} = \bar{x} \cdot \bar{y} = \bar{x} \,\square\, y = (1 \,\square\, x) \,\square\, y$.

9. (a)

c	a	b	c'	a'	b'
0	0	0	0	0	0
0	1	0	0	1	0
0	0	1	0	0	1
0	1	1	0	1	1
1	0	0	1	0	0
1	1	0	1	0	1
1	0	1	1	1	0
1	1	1	1	1	1

   (b) The inputs to the second Fredkin gate will be labelled a', b', c' and the outputs will be labelled a'', b'', c''. If $c = 0$ then $a'' = a$, $b'' = b$ and $c'' = c$. If $c = 1$ then, also, $a'' = a$, $b'' = b$ and $c'' = c$. Thus two such gates in series just give the identity function with outputs equal to the inputs. This means that Fredkin gates are *reversible*.

   (c) $a' = b \cdot c$. We can therefore construct and-gates.

   (d) $a' = \bar{c}$. We can therefore construct not-gates.

   (e) $a' = a + c$. We can therefore construct or-gates.

10. (a)

a	b	c	a'	b'	c'
1	1	1	1	1	0
1	1	0	1	1	1
1	0	1	1	0	1
1	0	0	1	0	0
0	1	1	0	1	1
0	1	0	0	1	0
0	0	1	0	0	1
0	0	0	0	0	0

(b) $a'' = a$, $b'' = b$ and $c'' = a'b' \oplus c' = ab \oplus (ab \oplus c) = c$, where we have used the fact that $\oplus$ is associative and $x \oplus x = 0$ (both of which should be checked). This means that Toffoli gates are *reversible*.

(c) $c' = a \cdot b$.

(d) $c' = \bar{a}$.

(e) $b' = b$, $c' = b$. This is fanout.

Exercises 2.4

1. (a)

(b)

(c)

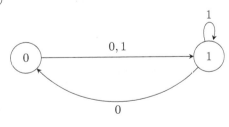

(d)

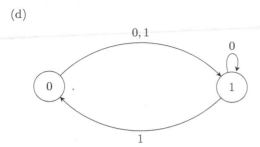

(e)

(f)

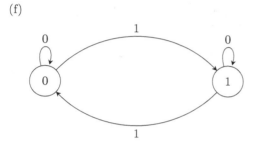

2. (a) The input/output table for the automaton is as follows:

y	x	y^+
0	0	0
0	1	1
1	0	0
1	1	1

(b) A Boolean recursion is $y^+ = x$.

3. (a) The input/output table for the automaton is as follows:

y_1	y_2	x	y_1^+	y_2^+
0	0	0	1	0
0	0	1	0	1
0	1	0	1	1
0	1	1	0	0
1	0	0	0	0
1	0	1	1	1
1	1	0	0	1
1	1	1	1	0

(b) The Boolean recursion is the following:

$$y_1^+ = \bar{y}_1 \cdot \bar{y}_2 \cdot \bar{x} + \bar{y}_1 \cdot y_2 \cdot \bar{x} + y_1 \cdot \bar{y}_2 \cdot x + y_1 \cdot y_2 \cdot x$$
$$y_2^+ = \bar{y}_1 \cdot \bar{y}_2 \cdot x + \bar{y}_1 \cdot y_2 \cdot \bar{x} + y_1 \cdot \bar{y}_2 \cdot x + y_1 \cdot y_2 \cdot \bar{x}.$$

Exercises 3.1

1. (a) Athelstan doesn't like himself.

    (b) If Athelstan likes himself then he isn't taller than himself.

    (c) Cenric isn't melancholy and he doesn't like Ethelgiva.

    (d) Athelstan is a cat if and only if either he is melancholy or he likes Ethelgiva.

    (e) Someone is taller than Cenric.

    (f) Athelstan and Cenric like everyone.

    (g) Athelstan and Cenric like everyone.

    (h) Someone is taller than Athelstan or someone is taller than Cenric.

    (i) Someone is taller than Athelstan or taller than Cenric.

    (j) All cats like Ethelgiva.

    (k) There is a cat that Ethelgiva doesn't like.

    (l) There is a cat that Ethelgiva doesn't like.

    (m) For each cat, either Cenric likes it or Ethelgiva likes it.

    (n) There is a melancholy cat taller than Cenric.

2. (a) $(\forall x)L(x, e)$.

    (b) $(\forall x)(L(c, x) \vee L(a, x))$.

    (c) $(\forall x)L(a, x) \vee (\forall x)L(c, x)$.

    (d) $(\exists x)(T(x, a) \wedge T(x, c))$.

    (e) $(\exists x)T(x, a) \wedge (\exists x)T(x, c)$.

    (f) $(\forall x)(C(x) \rightarrow L(e, x))$.

    (g) $(\forall x)(C(x) \rightarrow L(x, e))$.

    (h) $(\exists x)(C(x) \wedge L(e, x))$.

    (i) $\neg(\exists x)(C(x) \wedge L(e, x))$.

    (j) $(\forall x)(L(x, e) \rightarrow \neg C(x))$.

    (k) $\neg(\exists x)(L(x, e) \wedge C(x))$.

    (l) $(\exists x)(L(x, a) \wedge L(x, e))$.

    (m) $\neg(\exists x)(L(x, a) \wedge L(x, c))$.

3. In all cases, for every variable, regarded as a leaf, there is a path from that variable to a quantifier with that variable.

(a)

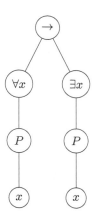

(b)

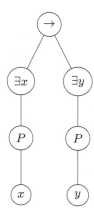

(c)

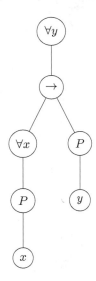

(d)

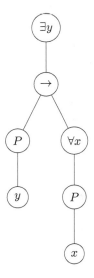

(e)

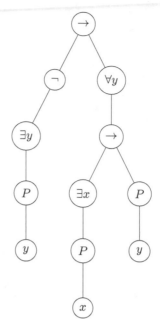

4. (a) $(\exists z)(P(z,x) \wedge P(z,y)) \wedge B(x)$.

    (b) $(\exists y)(\exists z)(P(x,z) \wedge P(z,y)) \wedge A(x)$.

    (c) $(\exists y)Q(x,y)$.

5. Directed graphs in which each vertex is coloured with exactly one of two colours.

Exercises 3.2

1.
$$(\forall x)R(x)*$$
$$\neg(\exists x)R(x)$$
$$|$$
$$(\forall x)\neg R(x)*$$
$$|$$
$$R(a)$$
$$|$$
$$\neg R(a) \; ✗$$

This argument is valid. Recall that all domains are assumed to be non-empty. Thus if $(\forall x)R(x)$ is true in an interpretation with domain D then there is an element $a \in D$ such that $\mathcal{R}(a)$ is true, where $\mathcal{R}$ is the interpretation of the predicate symbol R. It follows that $(\exists x)R(x)$ is also true in that interpretation.

2.
$$(\forall x)(\forall y)R(x,y)*$$
$$\neg(\forall x)R(x,x)\checkmark$$
$$|$$
$$(\exists x)\neg R(x,x)\checkmark$$
$$|$$
$$\neg R(a,a)$$
$$|$$
$$(\forall y)R(a,y)*$$
$$|$$
$$R(a,a) \; \textbf{✗}$$

All branches close and so the argument is valid.

3.
$$(\exists x)(\exists y)F(x,y)\checkmark$$
$$\neg(\exists y)(\exists x)F(x,y)\checkmark$$
$$|$$
$$(\forall y)(\forall x)\neg F(x,y)*$$
$$|$$
$$(\exists y)F(a,y)\checkmark$$
$$|$$
$$F(a,b)$$
$$|$$
$$(\forall x)\neg F(x,b)*$$
$$|$$
$$\neg F(a,b) \; \textbf{✗}$$

Since the truth tree closes the sentence is universally valid.

4. We prove that $(\forall x)(P(x) \wedge Q(x)) \models (\forall x)P(x) \wedge (\forall x)Q(x)$; the proof of $(\forall x)P(x) \wedge (\forall x)Q(x) \models (\forall x)(P(x) \wedge Q(x))$ is similar.

$$(\forall x)(P(x) \wedge Q(x))*$$
$$\neg((\forall x)P(x) \wedge (\forall x)Q(x))$$

$$\neg(\forall x)P(x) \qquad \neg(\forall x)Q(x)$$
$$| \qquad\qquad |$$
$$(\exists x)\neg P(x) \qquad (\exists x)\neg Q(x)$$
$$| \qquad\qquad |$$
$$\neg P(a) \qquad\qquad \neg Q(a)$$
$$| \qquad\qquad |$$
$$P(a) \wedge Q(a) \qquad P(a) \wedge Q(a)$$
$$| \qquad\qquad |$$
$$P(a) \qquad\qquad P(a)$$
$$Q(a)\textbf{✗} \qquad\qquad Q(a)\textbf{✗}$$

5. We prove that $(\exists x)P(x) \vee (\exists x)Q(x) \models (\exists x)(P(x) \vee Q(x))$; the proof of $(\exists x)(P(x) \vee Q(x)) \models (\exists x)P(x) \vee (\exists x)Q(x)$ is similar.

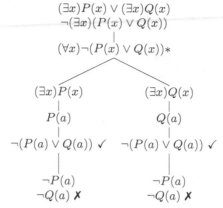

6. The truth tree for the first part of the question is as follows:

$$(\forall x)P(x) \vee (\forall x)Q(x) \checkmark$$
$$\neg(\forall x)(P(x) \vee Q(x)) \checkmark$$
$$|$$
$$(\exists x)\neg(P(x) \vee Q(x)) \checkmark$$
$$|$$
$$\neg(P(a) \vee Q(a)) \checkmark$$
$$|$$
$$\neg P(a)$$
$$\neg Q(a)$$

$$(\forall x)P(x)* \qquad (\forall x)Q(x)*$$
$$| \qquad\qquad |$$
$$P(a)\ ✗ \qquad\quad Q(a)\ ✗$$

Now consider the sentence

$$X = (\forall x)(P(x) \vee Q(x)) \rightarrow ((\forall x)P(x) \vee (\forall x)Q(x)).$$

For the domain of the interpretation take the natural numbers $\mathbb{N}$, interpret P as the subset of even numbers $\mathbb{E}$ and interpret Q as the subset of odd numbers $\mathbb{O}$. Then the sentence before $\rightarrow$ is true in this interpretation because every natural number is either odd or even. But the sentence after $\rightarrow$ is false because it says 'either every natural number is even or every natural number is odd', which is false. We have therefore found an interpretation in which X is false and so X is not universally valid.

7. We begin by proving that the sentence is universally valid.

$$\neg(\exists x)(D(x) \to (\forall y)D(y)) \checkmark$$
$$|$$
$$(\forall x)\neg(D(x) \to (\forall y)D(y))*$$
$$|$$
$$\neg(D(a) \to (\forall y)D(y)) \checkmark$$
$$|$$
$$D(a)$$
$$\neg(\forall y)D(y) \checkmark$$
$$|$$
$$(\exists y)\neg D(y) \checkmark$$
$$|$$
$$\neg D(b)$$
$$|$$
$$\neg(D(b) \to (\forall y)D(y)) \checkmark$$
$$|$$
$$D(b)$$
$$\neg(\forall y)D(y) \; ✗$$

In the given interpretation, the sentence says that there is someone who if they drink then everyone drinks. This sounds highly paradoxical. However $(\exists x)(D(x) \to (\forall y)D(y)) \equiv (\exists x)(\neg D(x) \lor (\forall y)D(y))$. But by an earlier question in these exercises, $(\exists x)(\neg D(x) \lor (\forall y)D(y)) \equiv (\exists x)\neg D(x) \lor (\exists x)(\forall y)D(y)$. However, $(\exists x)(\forall y)D(y) \equiv (\forall y)D(y)$. Thus the formula simply says that either everyone drinks or there is someone who doesn't drink which is not paradoxical at all. Refer back to Section 1.1 of Chapter 1 and my comment about translating between logic and everyday language.

8. *Reflexive.* This means that every vertex has a loop. *Irreflexive.* No vertex has a loop. *Symmetric.* If there is an arrow from vertex 1 to vertex 2 there also has to be a return arrow from vertex 2 to vertex 1. *Asymmetric.* If there is an arrow from vertex 1 to vertex 2 there is no arrow from vertex 2 to vertex 1. *Transitive.* If there is an arrow from vertex 1 to vertex 2, and an arrow from vertex 2 to vertex 3 there is also a direct arrow from vertex 1 to vertex 3.

(a)

$$(\forall x)(\forall y)(R(x,y) \to \neg R(y,x))*$$
$$\neg(\forall x)\neg R(x,x) \checkmark$$
$$|$$
$$(\exists x)R(x,x) \checkmark$$
$$|$$
$$R(a,a)$$
$$|$$
$$R(a,a) \to \neg R(a,a)$$

$$\neg R(a,a) \; ✗ \qquad \neg R(a,a) \; ✗$$

(b)

$$(\forall x)(\forall y)(\forall z)(R(x,y) \wedge R(y,z) \to R(x,z))*]$$
$$(\forall x)\neg R(x,x)*$$
$$\neg(\forall x)(\forall y)(R(x,y) \to \neg R(y,x))\checkmark$$
$$|$$
$$(\exists x)(\exists y)\neg(R(x,y) \to \neg R(y,x)) \checkmark$$
$$|$$
$$\neg(R(a,b) \to \neg R(b,a)) \checkmark$$
$$|$$
$$R(a,b)$$
$$R(b,a)$$
$$|$$
$$\neg R(a,a)$$
$$|$$
$$R(a,b) \wedge R(b,a) \to R(a,a)$$

$$\neg(R(a,b) \wedge R(b,a)) \checkmark \qquad R(a,a) \; \boldsymbol{\times}$$

$$\neg R(a,b) \; \boldsymbol{\times} \qquad \neg R(b,a) \; \boldsymbol{\times}$$

Bibliography

[1] S. Aaronson, *Quantum computing since Democritus*, CUP, 2013.

[2] M. Agrawal, N. Kayal, N. Saxena, PRIMES is in P, *Annals of Mathematics*, **160** (2004), 781–793.

[3] Marcus Aurelius, *The meditations*, translated by Robin Hard, OUP, 2011.

[4] M. Ben-Ari, *Mathematical logic for computer science*, Third Edition, Springer, 2012.

[5] A. Bierce, *The enlarged Devil's dictionary*, Penguin Books, 1989.

[6] G. Boole, *The laws of thought*, https://www.gutenberg.org/files/15114/15114-pdf.pdf, 1854.

[7] G. S. Boolos, J. P. Burgess, R. C. Jeffrey, *Computability and logic*, Fifth Edition, CUP, 2010.

[8] J. R. Büchi, *Finite automata, their algebras and grammars*, Springer, 2012.

[9] C. J. Colbourn, The complexity of completing partial Latin squares, *Discrete Applied Mathematics* **8** (1984), 25–30.

[10] R. Cori, D. Lascar, *Mathematical logic*, OUP, 2000.

[11] W. J. Dally, R. C. Harting, T. M. Aamodt, *Digital design using VHDL*, CUP, 2016.

[12] H. Delong, *A profile of mathematical logic*, Dover, 2004.

[13] A. Doxiadis, C. H. Papadimitriou, *Logicomix*, Bloomsbury, 2009.

[14] D. J. Eck, *The most complex machine*, A. K. Peters, Wellesley, Massachusetts, 1995.

[15] R. M. Exner, M. F. Rosskopf, *Logic in elementary mathematics*, Dover, 2011.

[16] T. L. Floyd, *Digital fundamentals*, Ninth Edition, Pearson Education International, 2006.

[17] R. W. Floyd, R. Beigel, *The language of machines*, Computer Science Press, 1994.

[18] J. H. Gallier, *Logic for computer science. Foundations of automatic theorem proving*, John Wiley & Sons, 1987.

[19] M. R. Garey, D. S. Johnson, *Computers and intractability. A guide to the theory of NP-completeness*, W. H. Freeman and Company, 1997.

[20] G. Gentzen, Untersuchungen über das logische Schließen. I, *Mathematische Zeitschrift* **39** (1933), 176–210.

[21] M. Gessen, *Perfect rigour*, Icon Books, 2011.

[22] J.-Y. Girard, *Proofs and types*, CUP, 2003.

[23] S. Givant, P. Halmos, *Introduction to Boolean algebras*, Springer, 2009.

[24] R. Goldstein, *Incompleteness: the proof and paradox of Kurt Gödel*, W. W. Norton & Company, 2006.

[25] T. Hailperin, Boole's algebra isn't Boolean algebra, *Mathematics Magazine* **54** (1981), 172–184.

[26] R. Hammack, *Book of proof*, VCU Mathematics Textbook Series, 2009. This book can be downloaded for free from `http://www.people.vcu.edu/~rhammack/BookOfProof/index.html`.

[27] D. Harel, *Computers Ltd. What they really can't do*, OUP, 2012.

[28] W. D. Hart, *The evolution of logic*, CUP, 2010.

[29] W. S. Hatcher, *The logical foundations of mathematics*, Pergamon Press, 1982.

[30] D. Hilbert, W. Ackermann, *Principles of mathematical logic*, Chelsea Publishing Company, New York, 1950.

[31] R. E. Hodel, *An introduction to mathematical logic*, Dover, 1995.

[32] A. Hodges, *Alan Turing: the enigma*, Vintage, 2014.

[33] D. R. Hofstadter, *Gödel, Escher, Bach: an eternal golden braid*, Basic Books, 1999.

[34] C. D. Hollings, *Scientific communication across the Iron Curtain*, Springer, 2016.

[35] E. V. Huntington, New sets of independent postulates for the algebra of logic, with special reference to Whitehead and Russell's Principia Mathematica, *Transactions of the American Mathematical Society* **35** (1933), 274–304.

[36] E. V. Huntington, Boolean algebra. A correction, *Transactions of the American Mathematical Society* **35** (1933), 557–558.

[37] M. R. A. Huth, M. D. Ryan, *Logic in computer science: modelling and reasoning about systems*, CUP, 2000.

[38] R. Jeffrey, *Formal logic. Its scope and limits*, Fourth edition (edited by J. P. Burgess), Hackett Publishing Company, Inc., 2004.

[39] R. W. Kaye, `http://web.mat.bham.ac.uk/R.W.Kaye/minesw/minesw.htm`.

[40] S. C. Kleene, *Mathematical logic*, Dover, 2002.

[41] W. Kneale, M. Kneale, *The development of logic*, Clarendon Press, Oxford, 1986.

[42] D. C. Kozen, *Automata and computability*, Springer, 1997.

[43] M. V. Lawson, *Finite automata*, Chapman & Hall/CRC, 2004.

[44] M. V. Lawson, *Algebra & geometry: an introduction to university mathematics*, CRC Press, 2016.

[45] G. Lemonde-Labrecque, Truth tree solver, `http://gablem.com/logic`.

[46] S. Lipschutz, M. Lipson, *Discrete mathematics*, Second Edition, McGraw-Hill, 1997.

[47] D. MacHale, *The life and work of George Boole*, Cork University Press, 2014.

[48] M. M. Mano, *Digital design*, Prentice-Hall International, 1984.

[49] B. Mates, *Stoic logic*, University of California Press, 1961.

[50] W. S. McCulloch, W. H. Pitts, A logical calculus of the ideas immanent in nervous activity, *The Bulletin of Mathematical Biophysics* **5** (1943), 115–133.

[51] G. H. Mealy, A method for synthesizing sequential circuits, *Bell System Technical Journal* **34** (1955), 1045-1079.

[52] E. Mendelson, *Introduction to mathematical logic*, Sixth Edition, CRC Press, 2015.

[53] A. A. Milne, *Winnie-the-Pooh*, Methuen, 1995.

[54] A. A. Milne, *The house at Pooh corner*, Methuen, 1995.

[55] R. Monk, *Ludwig Wittgenstein: the duty of genius*, Vintage, New Edition, 1991.

[56] R. Monk, *Biography of Bertrand Russell: 1872–1920 Volume 1*, Vintage, 1997.

[57] D. Mundici, *Logic: a brief course*, Springer, 2012.

[58] P. Nahin, *The logician and the engineer*, Princeton University Press, 2013.

[59] M. A. Nielsen, I. L. Chuang, *Quantum computation and quantum information*, CUP, 2002.

[60] C. H. Papadimitriou, *Computational complexity*, Addison Wesley Longman, 1994.

[61] C. Petzold, *Code: the hidden language of computer hardware and software*, Microsoft Press, 2000.

[62] C. Petzold, *The annotated Turing*, John Wiley & Sons, 2008.

[63] W. V. Quine, *Methods of logic*, Revised Edition, Holt, Rinehart and Winston, 1966.

[64] U. Schöning, *Logic for computer scientists*, Birkhäuser, 2008.

[65] D. Scott et al, *Notes on the formalization of logic: Parts I and II*, Oxford, July, 1981.

[66] P. Smith, *An introduction to formal logic*, CUP, 2011.

[67] R. M. Smullyan, *First-order logic*, Dover, 1995.

[68] R. M. Smullyan, *Logical labyrinths*, A. K. Peters, Ltd. 2009.

[69] R. M. Smullyan, *A beginner's guide to mathematical logic*, Dover, 2014.

[70] A. Sokal, J. Bricmont, *Intellectual impostures*, Profile Books, 1998.

[71] R. R. Stoll, *Set theory and logic*, W. H. Freeman and Company, 1961.

[72] A. Tarski, *Cardinal algebras*, OUP, 1949.

[73] P. Teller, *A modern formal logic primer*, Prentice Hall, 1989. This book can be downloaded for free from http://tellerprimer.ucdavis.edu/.

[74] N. Tennant, *Introductory philosophy: god, mind, world and logic*, Routledge, 2015.

[75] A. M. Turing, On computable numbers, with an application to the Entscheidungsproblem, *Proceedings of the London Mathematical Society* **42** (1937), 230–265.

[76] L. E. Turner, Truth table generator, http://turner.faculty.swau.edu/mathematics/materialslibrary/truth/.

[77] The Univalent Foundations Program, *Homotopy type theory: univalent foundations of mathematics*, `https://homotopytypetheory.org/book/`, Institute for Advanced Study, 2013.

[78] D. van Dalen, *Logic and structure*, Third Edition, Springer, 1997.

[79] L. Wittgenstein, *Tractatus Logico-Philosophicus*, Routledge, 2009.

[80] T. Yato, *Complexity and completeness of finding another solution and its applications to puzzles*, Master's Thesis, University of Tokyo, 2003.

[81] M. Zegarelli, *Logic for dummies*, Wiley Publishing, 2007.

Index

2-place predicate, 147
n-place predicate, 147
1-place predicate, 147

absorption, 25
Ackermann, Wilhelm, xii
adequate, 43
adjoining constants to a language,
    155
α-formula, 73
alphabet, 15
and, 3
and-gate, 121
Aristotle, xi
arity, 147
arrow, 148
associativity, 25
atomic formula, 147, 154
atomic statements, 9
atoms, 9

Berra, Yogi, ix
β-formula, 73
Bierce, Ambrose, xi
binary relation, 148
Boole, George, xii
Boolean algebra, 112
Boolean function, 121
Boolean operations, 105
Boolean recursions, 139
bound variable, 158
branch, 12
breadth-first search, 92

Cantor, Georg, 13
change of variable, 159
Church, Alonzo, xii

circuit, 121
combinational circuit, 120
commutativity, 25
compound statement, 10
concatenation, 15
conjunctive normal form, 50
constant, 146, 147
constants, 154
contingency, 21
contradiction, 21
Cook's theorem, 58
counterexample, 165
cut rule, 101

De Morgan's laws, 25
decision problem, xii
degree, 13
delay, 135
descendant, 12
directed graph, 148
disjoint, 106
disjunctive normal form, 49
disjunctive syllogism, 80
distributivity, 25
domain, 45
domain (of structure), 151
double negation, 25

edge, 11
element, 14
empty set, 14
Entscheidungsproblem, xii
existential quantifier $\exists$, 153

falsify, 21
fanout, 124
finite state acceptor, 40